Japan Houses in Ferroconcrete

ジャパン・ハウス：打放しコンクリート住宅の現在

ジャパン・ハウス

打放しコンクリート住宅の現在

編著——植田 実

写真——下村純一

Japan Houses in Ferroconcrete

coordinated by UYEDA Makoto

photographed by SHIMOMURA Junichi

目次[Contents]

固有な文体の饗宴

植田 実

■本書に収録されている安藤忠雄設計の細工谷の家は、関西テレビによって、建築家と施主との出会いから、家が完成するまでのプロセスを、克明に取材したドキュメントがつくられている。
この家が建つ、大阪の細工谷町は、ダンプカーだったら通れるか通れないかの狭い道に沿って商店が並ぶ。そこの古い棟割長屋の1軒分を切り取って、コンクリートを打ち込む工事風景は圧倒的で、空恐しい気さえする。しかし、ある意味でもっと印象的なのは、このような周辺の状況から、安藤がブロック造を推めたのに対して、施主のほうはあくまでもコンクリート造に固執し、結局は建築家が折れたかたちで、後者の実現に踏み切るという話し合いの場面である。
この場合、施主が希望したのは、性能ではなく、仕上げ材としてのコンクリートであることは間違いない。つまり打放しコンクリートでなければならないのである。現在のように優れた型枠大工が減ってきているうえに、住宅のような規模の建築でそれを行うことは、打放しコンクリートをますます高価にしているわけだし、この仕上げがそもそもきわめてハイスタイルの嗜好を代表するにもかかわらず、下町の商店街に住んできた人の理想的な家の形であるところに、打放しコンクリートと日本人の住まいの特異な関わりを見ることができる。しかも細工谷の家は決して特殊な例ではなく、他にもたこ焼屋さんや文房具屋さんが、雑誌で安藤の建築を知って、同じような打放しコンクリートの町屋を依頼し、実現させている。
一方、六甲や芦屋のきわめて恵まれた敷地に、大邸宅といえるスケールと住居水準をもつ住宅建築を、同じ打放しコンクリートで安藤が手掛けていることもよく知られている。関西だけでなく、東京にも同様の、安藤の大邸宅的建築をいくつか見ることができる。
ということは、日本の住宅は、施主の社会的地位や住宅が建つ場所の地域といった基準をもはやもち得ないことを示す。この基準の喪失は、他のすべての日本の建築家が手掛ける住宅においても当てはまる。さらには、あらゆる構法、あらゆる素材に関しても当てはまる特性である。
しかし、例えば鉄骨造、木造、そしてコンクリート造でも煉瓦やタイルや木で仕上げられた住宅は、それに加えて時代の状況、あるいはそれに対して、肯定的にせよ批判的にせよ、敏感であるとする建築家や住み手の意識というベクトルが働いている。ポストモダンの時代であり、多様性の時代であることを示す徴候は、とくに鉄骨造の軽快さに、あるいは逆に伝統的なものへの信頼を表す意志は、在来

Moteki House: ANDO Tadao, 1983 (photo: UYEDA Makoto)

Koshino House: ANDO Tadao, 1981 (photo: UYEDA Makoto)

The Ultimate in Style with Ingenuity

UYEDA Makoto

■Town house in Saikudani, designed by Architect Tadao Ando, was covered in a documentary film by Kanasai Television. This film describes all of the processes of its construction in minute detail. It proceeds from the encounter between the architect and the client up to the completion of the structure.
In the area where the house stands, Saikudani-cho in Osaka, there are shops on both sides of a narrow street, whose width is just about large enough for a dump truck. A scene during construction in which one of the old tenement houses is cut off for the concrete casting is overwhelming — even horrifying. But a more impressive scene, in a way, is the discussion between the architect and the client, Mr. Ando recommended a block building in consideration of the surroundings, but the client persisted in wanting a concrete building. Eventually the architect gave in.
It is clear that what the client expected from his house was concrete itself as a finishing material. It had to be architectural ferroconcrete. The recent decrease in the availability of good carpenters for concrete wood frame work makes ferroconcrete an extremely expensive finish for a small building like a private house. But concrete finishing represents an extremely fashionable taste, and people living in the downtown shopping districts regard houses of architectural ferroconcrete to be the ideal form of dwelling. This is an indication of the peculiar Japanese relationship between ferroconcrete and houses. The house at Saikudani is not a special example of this. After learning of Mr. Ando's building, even the owner of a street stand and stationery shop requested and built a similar townhouse of ferroconcrete.
At the same time, in the extremely favorable areas of Ashiya and Rokko, Mr. Ando has also built houses of grand scale and dwelling standards in a similar ferroconcrete finish. In Tokyo, there are also several grand looking houses of ferroconcrete by Mr. Ando.
In designing Japanese houses architects can no longer be primarily just concerned with the regional styles of the location and the owner's social status.
These criteria are of less importance than previously valued.
But even so, peoples still want houses which fit the circumstances of the time, regardless of whether they are steel, wood or concrete, or finished with bricks, tiles or wood. Signs show that we are in an era of post modern and diversity.
Traditional aspects can be seen particularly in the lightness of steel structure construction and the stability of conventional wood construction. But also, an oversensitivity to the circumstances of the time is commonly seen in both avant-garde houses and in conventional ones.
However, concrete buildings seems to

工法の木造の安定性に見受けられる。その意味では、いわゆる前衛的な住宅にも、保守的な住宅にも共通する、時代状況に対する過敏症が見られるのだ。
ところが打放しコンクリートの建築には、時代のどのような思想をも超えてしまう、奇妙な確信のようなものがつきまとっている。それはこの素材自身の抱えている本来的な問題のせいでもあるだろう。つまり、目にも肌にも暴力的に関わりかねない、住んでいくうえでの性能を含んだ問題、あるいは劣化という耐用年限に関わる難点を含んだ問題に対するために、建築家も住み手も半ば内側に向いた姿勢から由来するように思われる。造形上は完璧だとしても、半永久的に解決しない問題がつねに絡んでいる。それは、どんな単純な箱状のコンクリート住宅でも、可塑的な表現を存分に敢行した住宅でも、同じだ。
その結果、打放しコンクリートは時代以上に建築家の個性を驚くほど鮮やかに映し出す。形状のみならず、素材それ自体が、固有の文体として決定されてくる。それを技術論としてではなく、美学的感性的な側面から眺めていけば、木造建築の長い伝統のなかで、日本の住宅が出会ったコンクリートという新しい建築素材が、生活空間にどう関わり、建築家の創造性に何をもたらしたかが見えてくるに違いない。本書はその徴候群ともいうべき、代表的な建築家たちのここ4、5年の仕事を中心としたアンソロジーである。コンクリート造とは、ここでは打放しコンクリートに限っているのもそのためである。

■打放しコンクリートの住宅は、そのまま西欧風の住宅であることを意味しない。
アントニン・レーモンドが、1924年に東京・霊南坂に自邸を建てたとき、彼自身の言葉に従えば、内外ともに打放しコンクリートという点では、世界でもどこよりも早かった。彼は、フランク・ロイド・ライトが大谷石の装飾を基調とした旧帝国ホテルを通して認識した日本を、さらに深めた、つまりこの自邸によってライトを超えたと確信した。マヤ様式のようなものまで取り込んで、日本に対するヴィジョンを膨張させた(それはジョサイア・コンドルの、擬サラセン様式を加味した旧上野博物館と並ぶものであり、一方、日本人建築家の辰野金吾がイスラム様式を主調とした旧国技館とも表裏関係にある)態度に対して、日本の大工の伝統的手法のうちにある自然観という、いわばヴィジョンの消失点に、日本を見い出そうとした姿勢でもあった。
即ち、レーモンドは3つの禁止を自分に課している。サンドペーパー、釘、ペンキ。それはすべての素材を自然のままに残す、日本の建築の原則だった、と彼はいう。その形態も、中庭をコンクリートの塀で囲い(庭には「見越しの松」が2本植えられているだけ)、そのまま建築と一体化したマッスが自由に構成されて、要所々々に採光の開口部を控え目にとりながら、上階に向かってそのヴォリュームを減らしていき、最後には2本の煙突が空に残されている。西欧の重い邸宅建築のシルエットに比べれば、そこにも日本から学んだものが感じられる。レーモンドが求めたのは、飾り気のない、自然さと真実性だったのだろう。チェコスロヴァキアに生まれ、アメリカで設計活動を開始し、日本で人生の大半を過した建築家の仕事は、古い日本の建築をコンクリ

Architect's House: RAYMOND Antonin, 1923 (The Kentiku, October 1961)

hover contentedly outside the circumstances of the time. It might be caused by the fundamental nature of the material itself. It seems to derive from the half inward-looking attitude of the architects and dwellers. This is attributable to the problems of living performance which forcibly concern the eyes and skin of the dwellers. Or it may be due to the problem of durability. A combination of formative completion and incompleteness which can not be resolved appears equally in box-shaped concrete houses, no matter how simply they are built, and no matter how daringly changable the expression of the house is.
As a result, ferroconcrete reflects the vivid personality of the architect more strongly than the time. Not only its form, but the material itself has a style inherent to it. If you take a look from an aesthetic and sensitive point of view and not a technological one, how did the new building materials come to take part in living spaces? Also what kind of imagination did architects develop during the long period of traditional wooden construction? These should now become visible. This book is an anthology consisting mainly of the representative architecture during the last four to five years. Concrete construction is referred to as architectural ferroconcrete for the above reason.

■A ferroconcrete house with a bare finish does not by itself mean a Western house. When Antonin Raymond built his house at Reinanzaka in Tokyo in 1923, it was, according to his own words, the first concrete house with a bare finish ever constructed both inside and outside of Japan. Frank Lloyd Wright recognized this through his old Imperial Hotel ornamented by Oya stones. He was convinced that he could surpass Wright by building this house of his own. He took an attitude of expanding the vision for Japan, incorporating the mode of the Maya into his architecture (it ranked with Josiah Condor's old Ueno Museum flavored with pseudo-Saracenic style and, at the same time, it was related to the old Sumo wrestling Hall by Japanese architect Kingo Tatsuno, which used Islamic styles), it was also a quest to discover Japan in the view of nature found in the traditional method of Japanese carpentry. Raymond imposed upon himself three prohibitions: no sandpaper, no nails, and no paint. According to him, it was a principle of Japanese architecture to keep all materials in natural condition. With a coutyard surrounded by a concrete wall (there were only two pieces of "Mikoshi-no-matsu" in the garden), its form had a free composition unifying the courtyard into the architecture as it is, while providing lighting openings at important points. It reduces in volume as it goes towards the upper floor until there are only two pieces of chimney in the sky. When compared with the heavy silhouette of European residential architecture, its form shows us what he learned from Japan. What Raymond sought was unpretentious naturalism and reality. So the work of an American architect born in Chechoslovakia who spent most of his life in Japan greatly affected modern Japanese construction. He adapted old Japanese architecture to concrete and new wood construction. But at the present time, when our vision has been expanded and deepened, his consistent posture has become relativistic.
Relativity can also be pointed out from yet another perspective. If Raymond's own mansion were seen in a town alongside Western-style stone houses, its concrete walls may have seemed thin and

ートや新しい木造に翻案し、その典型をつくり出すことによって日本の現代建築に大きな影響を与えたが、現在のように再び建築のヴィジョンが拡大深化してきた時代には、彼の一貫した姿勢も相対化してきているといわざるをえない。
相対性は、また別の視点からもいえる。もしレーモンドの自邸を、西欧の石造の家が並ぶ町なかにおいて見れば、そのコンクリートの壁は薄くしなやかで、日本の建築のように思われたかもしれない。事実、鉄筋コンクリートは、積みあげられた石の壁に対立する膜面的な壁として、西欧の近代に現われたはずである。しかし、レーモンドのコンクリートの家は、竣工当時の写真に見られるように、日本の伝統的な木造の家並みのなかに割り込んで、木に対する石の強さを印象づけている。つまり石造を代用するものとしてのコンクリートというイメージは、現代に至るまで日本人にとっては変わっていないのである。
しかし、彼の霊南坂の自邸のなかで今も日本人の住まいの特性として生きのびてきているものがあるとすれば、それは仕上げられない、素材のままの壁と天井である。それは、過剰な装飾を一切剥ぎとろうとした、西欧のモダニズム初期の住宅とも異なっている。
繰り返すが、打放しコンクリートの壁は、日本の住宅の洋風化を意味しない。大邸宅であれ、町なかの都市型の小住宅であれ、ほとんど例外なく、今でも日本の住宅は入口で靴を脱いで上がる場である。畳や伝統的な板の間と変わらない部屋に、ファーニッシュしない壁としてコンクリートが顔を出しているだけであり、その意味ではレーモンドの洞察どおりに、日本の住まいは過去からの連続性を保っている。
むしろ、その衝撃的な取り合わせは、戦後、日本の建築家の仕事が小住宅、つまり中流階級を対象として集中しはじめてからより一層積極的に意識されてきた。肯定的な意味合いをもって、当時の一部の建築家から「汚な仕上げ」などと呼ばれていたコンクリートの肌は、零度のファーニッシングへの意欲を、むしろかき立てたのだった。こうした、見方によってはシニカルな、しかし他にこれ以上の解決を望まない美意識の回路は、西欧の室内のコピーから逸脱して、日本独特の建築言語へと偏り、定着してくる。
最近は、驚くほどの大邸宅を、日本で散見するようになったが、こうした家ではとりわけ室内の装飾的要素は極度に抑制されている。一方、一般的な住宅においては、室内風景はほとんど手つかずのまま、気楽な雑居的なモノの氾濫にまかせている場合が少なくない。ファーニッシングの意識は、建築そのものに比べて未成熟であるといえるし、日本人の生活空間として相応いし姿を模索しているとも思えるのである。

■戦後の小住宅設計の歴史のなかでは、1950年前半に、既にもっとも正統的に、コンクリート造、小規模の住まい、西欧と日本という問題を、絶妙なバランスで描き出した例が出現している。清家清の「私の家」である。
10m×5m、一部地階、平屋建ての実に小さな住宅だが、東西の妻壁と、中央の最少限の耐震壁と鉄骨トラスが天秤のように屋根やスラブを支えているだけで、内部の間仕切りやドア(便所さえも)を一切取り払った、シンプル

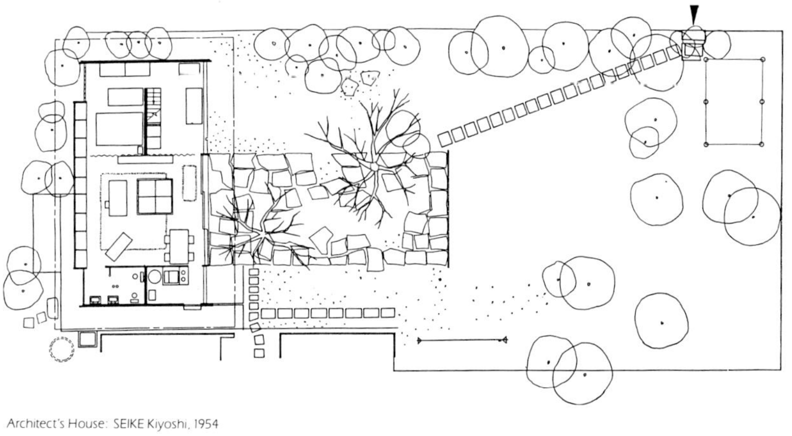

Architect's House: SEIKE Kiyoshi, 1954

Architect's House: SEIKE Kiyoshi, 1954 (photo: UYEDA Makoto)

pliable, giving somewhat of the feel of Japanese architecture. In actuality, ferroconcrete may well be said to have emerged in modern times in the West as a membranous wall in relation to the piled stone wall. Nevertheless, Raymond's concrete house gives one the impression of delivering the strength of stone relative to wood by breaking in on rows of traditional Japanese wooden houses, as shown in the photo at the time of its completion. In other words, the Japanese people have steadfastly embraced the image, without change to the present age, of concrete as a substitute for stonework.
If, in his house at Reinanzaka, there is something still surviving which matches the life of present Japanese, this can be found in the unfinished bare wall and ceiling. This is dis-similar to earlier European modernist houses which took off all excessive ornaments.
So the unfinished concrete wall does not mean the Westernization of Japanese houses. Whether it is a grand house or a smaller city house in town, the entrance of a house is a place where, almost without exception, people take off their shoes. It is an unchanged room with the traditional "tatami" mat. It has a bare board floor and unfinished ferroconcrete walls. In this sense, Japanese houses maintain continuity from the past, as Raymond discerned. This charged combination has become more positively recognized since Japanese architecture began to center upon smaller houses, namely, to those of the middle class after World War II. Concrete was called "a dirty finish" by some of the architects in those days. This had a positive meaning, which came from their volition for zero furnishing. Thus, there is now a consciousness of beauty which does not wish any better solution than this. Cynically, it may appear to have started as a deviation from copying European interiors. But it has now formed an architectural language peculiar to the Japanese.
Surprisingly grand looking houses have been built lately, and in these houses, decorative elements in particular are extremely restrained. At the same time, in general houses, there are many cases where the interior is left almost untouched, leaving it to be flooded with articles of everyday life. In comparison with construction itself, furnishing consciousness has not ripened yet. This is also important in the search for a suitable living space for Japanese.

■In the design history of small houses after the war, there appeared a modern orthodox small house made of concrete. This represented Europe and Japan in an exquisitly balanced way. This had already appeared in the early 1950. An example is "My house" by Mr. Kiyoshi Seike.
This is a tiny one-story house of 10m X 5m with a partial basement. It has a transverse wall from east to west, a minimal earthquake-resistant wall at its center, and a steel truss supporting the roof slab like a balance. There are no interior partitions or doors (not even a toilet door): all is one living space. Even a concrete wall appears as if it were "erected" like a print block. (Mr. Seike has always had awareness of form which contrasts "construct" and "erect".) It is close to what we think of as a Japanese house, but he tried a European-style living approach in its living content. Flat black stones were laid on the floor and the dwellers live with their shoes on. The floor continues to a spacious outside garden. A movable "tatami" mat of 150cm square with casters complements the stone floor by

な一室住居である。コンクリートの壁さえも版のように「建て起こ」したような見せ方をしている。(清家は、建築における「築く」ものと「建てる」ものとの対照的形式をつねに意識していた)いわば日本の建具を中心にした家の構成に近いともいえるが、一方、生活の内容としては西欧の住宅の住み方を試みた。鉄平石を貼りつめた床に、靴のまま生活することである。その床は、屋外の広い庭に連続している。この石の床を補って、車のついた150cm角の可動畳を、生活の日本的部分――着物をたたんだり、お茶を飲んだりするための最少限の床として用意している。
コンクリートの壁は、外側に面しては煉瓦仕上げ、内側では人造石小叩き仕上げ。妥当で穏やかな解決をはかり、打放しのままの面は天井だけである。にもかかわらず、架構からディテールまで曖昧な部分を一切残さぬ精神は、その時点であまりにも完全な解答を出してしまったとさえいえる。
この住宅を継承した例はない。もしあるとする場合、この住宅に対する見方は、2つしかない。ひとつは、正確には一室住居ではないという見方である。家の中央の耐力壁を極限にまで切りつめた壁の形ではなく、間取りに対する整合的な架構のために発生しつつある壁であるとする見方である。この解釈からすれば、間取りと架構の整合性をさらに成長されることができる――すなわち、数限りなくヴァリアントがつくり出される。しかもその枠を、木造や鉄骨造まで拡げることができる。このヴァリアントは、現在でも限りなくつくり出されつつある。しかし、この「継承」の見方のなかでは、清家の住宅という原型は、当然見失われている。
もうひとつは、極点に達した最小限の一室住居であるとする見方である。この軸を動かさずに、清家の「私の家」に続く住宅を探そうとすると見えてくる例はごくわずかである。架構と間取りに関する、全く新しい思想と発想を待たねばならないからだ。

■戦後の日本の小住宅の展開は、プランニングの展開そのものだろう。それは例えば、西海岸のケーススタディハウスを典型とする、アメリカの核家族の標準住宅をモデルとした、新しい家族像のための合理化された住宅は、平面計画においてすべてが語られ、そこには架構の整合性もすべて映し込まれていたといっていい。この流れを追っていくと、1958年に突然、全く特異なプランに出会う。もちろん、菊竹清訓の「スカイハウス」である。
1辺7.2mの正方形の完全な一室空間。その四周を1.2mの回廊がめぐらされている。この回廊のどこにでも、必要に応じてムーブネットとしての洗面・浴室やキッチンのブースが置かれ、収納もムーブネットとして、居室のなかに自由に配置される。こうした水まわりのコアがしばしば架構と抱き合わされて構成されていたプランニングが、この住宅では、何もない生活空間を揺らぎない主体として、設備・収納等をサブ・システムとして部屋の外に放出しているのである。そして、さらにその外側にはじめて架構の壁が出現し、家全体を抱いている。
この菊竹の自邸が、プランニングにおいて新しい発想を見せているだけでなく、家全体を4本の壁柱で持ち上げている架構形式においても全く独自であるとことは、いまさら言及

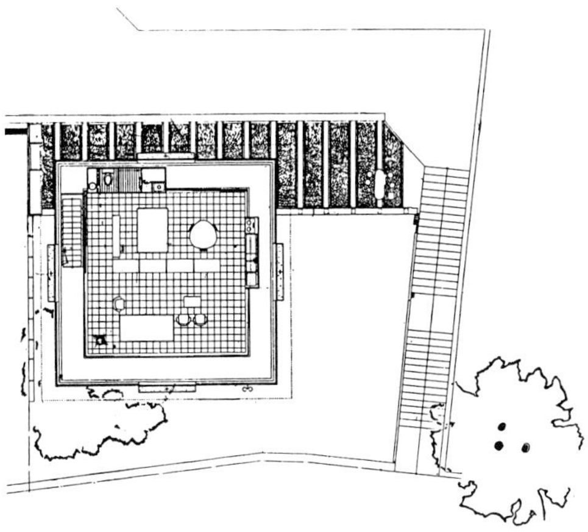

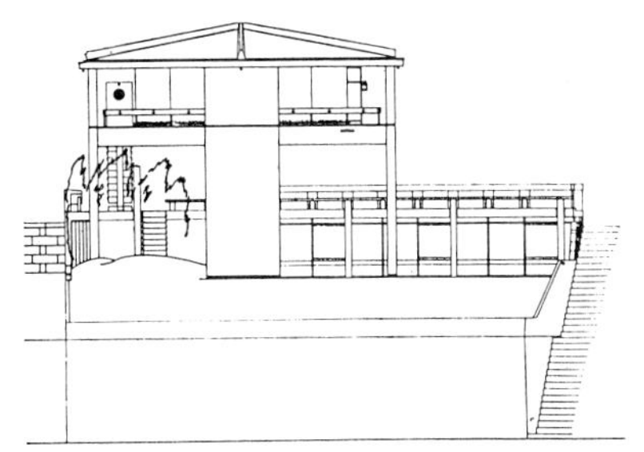

Sky House: KIKUTAKE Kiyonori, 1958

providing a minimal Japanese surface for folding kimono or enjoying tea.
A concrete wall with a brick finish from the outside and an artificial stone dabbed finish from the inside presents a reasonable and mild solution. It is only the ceiling which remains bare and unfinished. His desire to leave nothing vague throughout, from the frame to the details, may have rendered much too completed a solution at that point.
There is no house to have succeeded this example. If there is a reason, it is that the house is not a one room living space in its true meaning. We can look at the central earthquake resistant wall not as a wall curtailed as much as possible, but as a wall originating from the frame for partitioning. From a comprehension as this, conformity between the partition and the frame can grow further: in other words, innumerable variants can be created. Moreover, this frame work can be extended to wood and steel structures. And these variants can be made in countless numbers. But from the view point of "inheritance", the originality of Mr. Seike's house is naturally lost. Another way we can view this house is as a one-room living space minimized as completely as possible. Only a few examples can be found following Mr. Seike's "My house" when you look for one without deviating from this view point. You may need a new thought and a new generation.

■The development of small houses in Japan after the war might be said to be the development of planning itself. For example, a rationalized house for a new family image modelled upon standard house for an American nuclear family can be seen in a typical case study house on the West Coast. All about the house is told in the floor plan, where conformity of the frame is also reflected upon. When this flow is pursued, we suddenly encounter a totally unique plan in 1958. This is "Sky house" by Mr. Kiyonori Kikutake. A perfect one room living space in the shape of a square has a length of 7.2m on one side. The room is surrounded by a corridor of 1.2m in width all around. Booths for the wash basin, bathroom, and kitchen can be placed anywhere as necessary around the corridor. Storage can also be freely positioned any place within the living space. These water facilities often constitute combinations with the frame, but in this house, the empty living space is the main subject, and facilities and storage, are subsystems. This brings the framing wall to our eyes for the first time.
This house of Mr. Kikutake not only displays new concepts in planning, but also shows uniqueness in the frame structure: the whole house is supported by four wall columns. It is clear that the concrete is the only material in this house which can not be replaced by another one. A wall column has civil engineering characteristics. It is not unified into the structure of the house, but appears as an inevitable frame. The living space central to the house is distinct from the wall and is surrounded by the fixtures of the house.
A square vacant space has an original form like a shrine, and is linked with ancient Japanese living. Further, the house is designed mainly for a couple; a children's room and an atelier cluster around it rather reluctantly. This is not simply a house of extremity; it is a realization of a theoretical Metabolism advocated by Mr. Kikutake and his associates. It is a perfect model for furniture, construction, and city. This dwelling, which converges with its ancient origins, is not only a coherent explanation of problems from furnitures

する必要もないだろう。同時にこの時点で、打放しコンクリートが、この住宅にとって他に代えることのできない唯一の素材として姿を見せた。土木的な性格をも併せもった壁柱は、家そのものの構造と一体化せず、しかし必然的な架構として現れている。しかも、中核の居室そのものは壁から離れて、建具だけで囲われている。
正方形のがらんどうの空間は、神殿のようにあまりにも原型的だが、遠い古いところで、日本の住まいと結びついている。さらに夫婦2人の住まいとして構想され、子供部屋やアトリエは、その周りにとりつく形が否応なく明示されている。これは単に極限の住宅というばかりではなく、菊竹たちの提唱するメタボリズムの理論の実現——つまり家具・建築・都市を貫徹するモデルの提示としても完璧であるといえるだろう。この住宅は、原形への収斂、家具から都市の問題までを一貫する説明、住まいであると同時に住宅を超越するものをひとつの建築のなかに見せるという衝動を、後続する若い世代に植えつけずにはおかなかった。「スカイハウス」には、新しい時代の住宅の内容——つまり家族という中心を問い続けて、ついにはそこから「虚」がにじみ出るに至った様相が見られる。神殿のように見えてくるのである。西欧の目から見れば、あまりに概念的抽象的に見えるかも知れない家族の姿が、われわれには不思議なリアリティを帯びているのだ。そして、この虚の磁場は、生活と一体になろうとする架構を遠ざける。コンクリートの壁柱は、住居から隔てられ、同時に、住居をこの上なく明快に可視化している。

■1960年代に入って現れた、もうひとつの正方形プランの住宅は、菊竹の自邸を引き継ぐというよりは、「スカイハウス」と表裏の関係で読めるような、相反と一致がある。
大分市内に建つ、磯崎新のN邸は、「スカイハウス」の健康な明快さを反転したような曖昧性を、これもきわめて論理的に示している。やはり地上から持ちあげられた正方形プランの住宅だが、前者の壁柱が天高く、楽々と住居を包んでいるのと対照的に、ピロティの脚は細く短くなって住居の四隅に立てられ、その重みにやっと耐えているといった風情で、住居の箱の外側にズレ出している。住居は約12m角、四隅にコンクリートの立方体を置き、その間をガラスブロックの壁でつないでいる。この形をそのまま縮小したような、4つの立方体からなるモニター(天窓)が屋根の中央部に立ち上がる。開放を嫌う閉じた箱である。ここで目指されているのは、架構と生活の構成——プランニングとの全くの「無関係」である。それは大屋根の下のユニヴァーサル・スペースといった関係でもない。架構は独自にひとつの建築的形式を完結する。そして、四隅の立方体に穿たれた銃眼のようなガラスの固定窓、ガラスブロックの壁、モニターといった同一正方形パターンの反復とそこからの採光は、部屋々々に対して建築計画的に配慮されたものではない。一室住居のなかは大小の収納間仕切と家具だけで構成されているが、固定した開口部からの光は、家具も空間も巨大な静物のように見せる。磯崎自身の説明によれば、それはフェルメールの室内に注ぎ込まれている光を想定したという。つまり採光計画をイリュージョンの問題に転換している

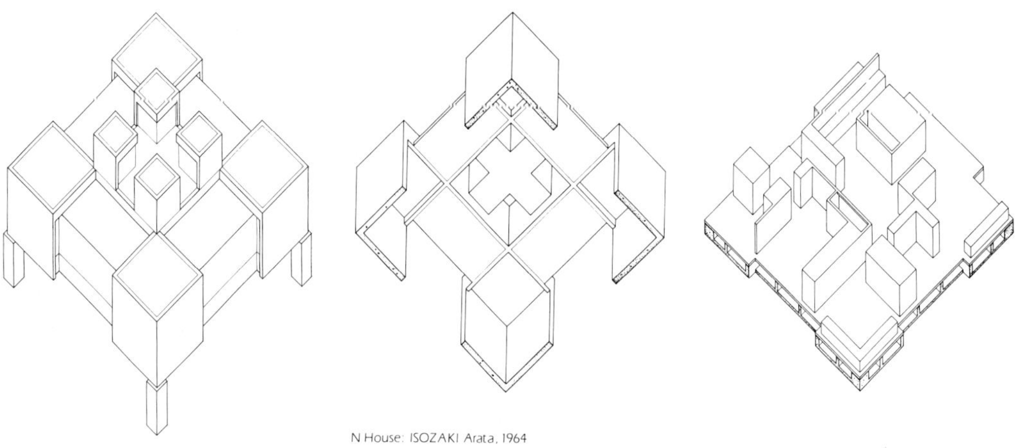

N House: ISOZAKI Arata, 1964

to cities, if invites the younger generation to follow its impetus and present in a building something surpassing housing. This has become, a fundamental attitude of Japanese architecture up to the present. "Sky house" delivers aspects of its expression in "hollowness", which is the ultimate result of ongoing scrutiny of the core content of the house centering on the family. Thus aspects are brought forth that may be likened to those of a shrine. The magnetic field for such emptiness wards off construction work designed to achieve integration with living. Concrete walls and pillars are kept away from the house but at the same time they play the role of presenting the house in an explicitly visible manner without comparison to extraneous elements. To the eyes of the Western world, the status of the family may be sonewhat conceptual and abstract. In Japan, however, we see it as a source of wondrous reality.

■Another square house which appeared at the beginning of 1960 does not exactly following the example of Kikutake's house; rather, it is in the relative relationship between a minus pole and a plus pole to "Sky house".
Mr. N.'s house by Mr. Arata Isozaki which stands in Oita City represents the reversed ambiguity of the healthy clarity of "Sky house" with theoretically extreme clarity. This house is also lifted up from the ground, but while Sky house's wall column is tall, lifting the ceiling high above and comfortably enveloping living, Mr. N.'s house, in contrast to this, has short and thin-legged pilotis at four corners. It appears that the house is finding it hard to withstand its weight and the pillars are protruding out of the box. The house is about 12m square and cubes of concrete are placed at four corners. Walls made of glass blocks separate these cubes. A top light consisting of four cubes, looking exactly like a contraction of the house, stands upward at the center of the roof. It is a closed box that does not want to open. Aimed at here is a composition of frame and life... "No relationship" planning. It is not in relationship with the universal space under the great roof. The frame itself is self-completing as a construction style. Nothing is deliberately planned: not the window with glass panes fixed at the cubes of the four corners like loopholes, not the wall made of glass blocks, not the monitor on the roof top representing a repetetion of the square pattern, and not the lighting which comes from it. Inside one room, living space consists of storage partitions and furniture only, but light coming from fixed opening makes the furniture and the space look like a gigantic still life. According to an explanation by Isozaki himself, he assumed light coming inside the room of Vermeer. In other words, the lighting scheme is converted to a matter of illusion.
"Light and dark", and the superiority of illusion which hang around architecture by Isozaki react strongly against such a small house. This has succeeded in derailing the design flow of Japanese houses. It is not a "general solution," and away from the circumstances of the times it will affect the young generation like a poison after 1970. At present, towards the end of the 1980, it has infiltrated so deeply that the word vulgarized might be appropriate for it; however, Isozaki's unique sense for architectural concrete which was established at that time might be his own even now. At the time this house appeared, his queer, heavy proportions and figures might not have been accepted as gallant and beautiful. Isozaki's masse of

のである。
「光と闇」という、磯崎の建築に終始つきまとっている幻覚性の優位が、これほどの小住宅においても強く作用し、日本の住宅計画の流れから独立することに成功しているわけで、「一般解」でないことと併せて、時代状況から隔たって、70年代以降に出現する若い世代に、毒のように効いていく。それは80年代末期の現在、風俗化しているとさえいえる浸透ぶりだが、このときすでに確立している、打放しコンクリートを扱うときの磯崎の独自の感覚は、今に至るまで磯崎ひとりのものだろう。この住宅が出現した当時は、尚さら、彼の奇妙に重いプロポーションや造形は、美しく颯爽としているとは見られなかっただろう。現在、磯崎のこのコンクリートのマッスはますます斬新であり、ビルの谷間に埋もれた、ぴっちり閉じた形も陳腐化していない。
比較的最近の磯崎の住宅、すなわち70年代末期から80年代はじめにかけて実現した一連の住宅は、いずれもコンクリートの箱の上にシリンダーをのせたシルエットが目立つ。そのうちの2軒――東京の貝島邸と福岡の林邸は、芝生の上に白く輝くヴィッラである。前者はとくに、狭小な敷地のなかで、イタリアン・ヴィッラの形式を、純粋立体に抽象化しつつ踏襲している。後者はその正面性が維持されたまま、必要な規模だけ建物が長く引き伸ばされ、ベルベデーレのあるロマンチックな庭園が凝縮してつくられている。次の2軒は、アートのコレクターとアーティストのための、黒いギャラリーと住居である。前者は、日本海側の雪国のなかで、プライヴェートなギャラリーを抱き込んだ住宅で、ほとんど全面を打放しコンクリートで徹底した閉じた室内に、コンクリートの箱とヴォールトの組み合わせが、立方体な光を投げ入れている。後者は、東京の町なかにギャラリーとアトリエと住居とを重ねた、メタル・アーティストのための都市型住宅である。最上階にある居間は、コンクリートを厚くくりぬいた、寺院のようにも思える暗い室内であり、照明もより一層、闇を招くかのように抑えられたデザインで、その部屋の底で金属や石の家具・什器が微光を放っている。
白いコンクリートは西欧の光をとらえ、黒いコンクリートは日本の闇に同化することを意識しているのだろう。ここでも「光と闇」のイリュージョンが、コンクリートの力を借りて持続され展開されている。この幻覚性は、家を家の形のレベルに収めない。家から捨てられた建築、例えば廃墟や機械のなかに住むような空間を暗示してくる。

■清家清と同世代の吉阪隆正は、清家の「私の家」の翌年に、やはりひとつの実験といっていい自邸を建てている。
その最初のスケッチは、吉阪がル・コルビュジェの下で働いていた、パリから家族への手紙のなかに「日本で建てるための新材料の家」として登場する。ピロティに持ちあげられた数層の部屋と、屋上に拡げられた傘のような屋根。同時期のコルのジョーダン邸を連想させるシルエットであり、この自邸の実現と、そのすぐあとに続く、打放しコンクリートの可塑的形態が、当時の日本には先例のない強烈さで出現した「ヴィラ・クゥクゥ」がやはりロンシャン礼拝堂とほぼ同じ年代であることは、コルビュジェアンとしての吉阪を印象づけず

Architect's House: YOSHIZAKA Takamasa, 1955 (photo: UYEDA Makoto)

Misawa House: YOSHIZAKA Takamasa, 1974 (photo: UYEDA Makoto)

concrete structure is all the more novel now. It's slightly closed appearance found rarely among buildings is still not hackneyed.
A series of Mr. Isozaki's comparatively recent houses built from the end of 1970 to early 1980 attract attention by the cylindrical silhouette of their boxes. Two houses among them, one of Mr. Kaijima house in Tokyo, and the other of Mr. Hayashi in Fukuoka are villas shining white on their lawns. Particularly the former is an abstracted Italian villa in a pure cubic style with a narrow land area. The latter, with its front appearance well maintained, extends far backwards and a romantic garden with observatory is densely constructed. Two other houses are a dark gallery for an art collector and a house for an artist. The former is a house with a private gallery in snowy country facing the Japan Sea. It is built of almost all bare concrete and cubic lighting is thrown inside the thoroughly closed rooms through a combination concrete box and vault. The latter is a city house in a downtown area of Tokyo for a metal artist. It is a gallery, atelier, and house for living. Living space on top floor is a tall hollow of deep concrete. It looks like a dark temple, with lighting designed so that it invites further darkness. Furniture and utensils of metals and stones glimmer at the bottom of the room.
Perhaps white concrete is aware of catching Western lighting while dark concrete is adapted to Japanese darkness. Here the illusion of "Light and darkness" is supported by the power of concrete. This illusion does not confine a house to the level of a house. It implies a space to live, conjuring visions of ruins and machinery, forms of architecture far away from the idea of a house.

■Mr. Takamasa Yoshizaka, who is of the same generation as Mr. Kiyoshi Seike, also built his own experimental house in the year following the construction of Mr. Seike's "My house."
His first sketch of the house was in a letter to his family from Paris where he worked for Le Corbusier. He called it, "A house of new materials to be built in Japan." Rooms on several floors rise from the ground on stilts under a roof that looks like a half-opened umbrella. The silhouette is similar to that of the Corbusier's Jorhdan house, which was built at about the same time. Also, "Villa Cou Cou," which was built of unfinished concrete appeared immediately after the finish of Mr. Yoshizaka's house, creating a strong impression on Japanese architects. That it appeared at almost the same time as the Longchamp Chapel further added to its effect on people as a Corbusierien. Corbusier's form and its rough texture of unfinished concrete were accepted at the time as being architecturally in the middle of the road. But there is more to these two houses than just being Corbusierien. Above all else, his house consisted only of a floor and pillars, built on artificially made ground. His friends who lost their houses during the war used to gather together on his premises and built shelters for themselves within his compound. Under these circumstances, their very first step was to make a floor; they say how that these structures had neither walls nor roofs. Given such primitive living conditions, the next step was to pour concrete. In much the same way as Mr. Seike's house having a flat black stones floor to be dwelled in with shoes on was an experiment, Yoshizaka's house was also entered with shoes on and a complementary bench with "tatami" mat and alcove were provided. These two exam-

にはおかなかった。そこでは、コルの形態と、荒々しい打放しコンクリートのテクスチャは不離不即のものとして受け入れられた。
しかし、この2軒の住宅には、単なるコルビュジェアンの枠を超えたものがある。自邸には何よりも床と柱だけがコンクリートでつくられた(壁はブロック積み)、人工土地である。その敷地には戦災によって家を失った知人たちが集まってきて、てんでにバラックをつくって住みこんでいたのだが、その家づくりの第一歩は床づくりであり、屋根も壁もない時期もあったという。そうしたプリミティブな生活の基盤を次のステップに進むために、コンクリートが流し込まれたに違いない。清家の家が鉄平石の床に靴のままという実験であったように、吉阪邸も靴のまま室内に入り、その補完として畳のベンチやアルコーブが用意された(2軒とも結局は家の中では靴を脱ぐ生活に戻るが)のも、きわめて自然な展開だった。
同時に、吉阪のコンクリートへの信頼は、「私はなぜコンクリートにこだわるのか」という一文で、端的にその原風景を語っている。それは、中国のモンゴル自治区の草原で見た「燕の巣みたいな」泥造りの家であり、サルディニア島の、石の上から漆喰と石灰でぬりたくって「ヌルッとした一つの塊」になった「大きな抽象彫刻」だった。住宅計画の合理化が大テーマであった時代に、それは早すぎるともいえる原風景の提示である。しかも、西欧と日本という構図からも逸脱して、コンクリートは地球そのものの表象であるような視座がとられている。
コンクリートは石や砂を主体とし、つまり母なる大地にもっとも近い感触だから、人工土地になりうるのだ、と吉阪はいう。それはピロティで空中に持ち上げられるが、同時に地底の住居でもあるという二重性を語っている。吉阪の住宅はどれも目鼻立ちがはっきりしているために壁の性格が強く感じられるが、実は壁はない。もぐらのように自在に穴を掘っていった結果が、最後に土を払って空中に露出した住まいの形である。それは彼の最後の住宅の仕事にまで維持され、そのプロセスでコルの原則から離れて、吉阪独自のヴィジョンをあらわに見せてくる。

■60年代後半に出現した東孝光の自邸も「塔の家」という名称を与えられながら、実は地底の家――いわば垂直の洞穴である。吉阪邸がそうだったように、とりあえずの住む場所の確保が内外とも打放しコンクリートのまま構造体によって達成され、結局は20年間そのままで住まわれてきた。洞穴、すなわち、これも住宅計画の原則を一切無視した、階段室だけともいえる一室住居であり、清家の家と同様、トイレにもドアはない。そこから6層の床が発生している。この多層空間には、架構と生活構造との整合・非整合といった判断基準を超越してしまった、全くの自由がある。ある意味ではとりとめない形であり、確実なのは東のいう「意外に暖か味のある、優しい表情をもっている」コンクリート打放しの肌と空間なのだ。そこで展開されている日本人の住まいの光景は、30年以上経った、清家の「私の家」の現在とよく似ている。一見雑然としているが、建築の物質性との信頼関係のなかで育まれてきた私的聖域である。
戦前の日本において、モダニズム初期の住宅、すなわち国際様式の住宅が、そのスタイルを

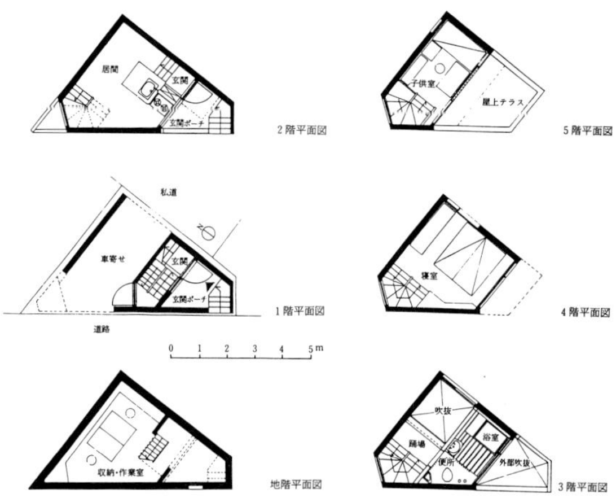

Tower House: AZUMA Takamitsu, 1967(photo: UYEDA Makoto)

ples must have developed quite naturally. (Eventually, the two of them returned to the ordinary Japanese way of living, which is to take shoes off at the entrance.) Mr. Yoshizaka's reliance on concrete is frankly explained his essay, "Why I stick to concrete." He compares this type of dwelling to a "Swallow's nest" made of mud in Monogolia, as well as to a type of housing found on Saljinian Island, which is made of plastered and lime-covered stone, has a "slimy" surface, and looks like "a giant abstract sculpture." As an original form for living, this proposal was too early, coming as it did at a time when rationalization of house design was a major theme. In addition, his view deviated from the composition of Europe and Japan houses and considered concrete as a representation of the earth itself.
"Since concrete is made of stones and sand, it has a feeling of being close to Mother Earth, and so it makes a good artificial ground," Mr. Yoshizaka says. It has the dual characteristics of being used as stilts to lift a structure into the air and, at the same time, it can create basement dwelling. Since all of Mr. Yoshizaka's houses are finely chiseled, a feeling of having walls is strong, but, in fact, there are none. This results from envisioning a form as a tunnel dug by a mole and then exposed to the air – the form remaining is the house. This concept was maintained into his house project, and during the course of his development, he evolved positively from Corbusier to create his own vision.

■The house which Mr. Takamitsu Azuma built in the late 1960s was called "Tower house," but it was, in fact, a basement house or, so to speak, a perpendicular cave. As was the case with Yoshizaka, Mr. Azuma had to have a house to live in temporarily and realized it in a structure of concrete with bare finish both outside and inside – in which he eventually lived for 20 years. The cave, which shows a complete disregard of basic house planning, consisted of almost of no other facility than a stair unit, and had no door for the toilet, similary to Mr. Seike's house. From there six floors were formed. In this multi-floored space, such criteria as the structure's conformity or nonconformity with frame and living requirements were suspended, and a transcending freedom achieved. In a way, it is a rambling form. And it is also certain, according to Mr. Azuma, that its ferroconcrete finished living space has an "unexpectedly warm and tender expression." It resembles the form of Japanese house developed some thirty years ago in Mr. Seike's "My house." Though it may appear pellmell, it is a private sancturay created under the reliable relationships within architectural materialism.
In prewar Japan, houses in the early stages of modernism or houses patterned on the style of Western architecture, gave foremost importance to their outer appearance, especially to the form of the roof, which constituted the decisive factor for such houses to attain their own style. Although, inside the building, the builders adopted the salon format characterized by huge volume and chair-type flooring, there were few cases in which conventional Japanese living spaces had to stand face to face with them in comparison. The greater majority of these were large house with exteriors looking like white cubcs of sugar, and were equipped with a large vestibule, tatami floor-matted rooms and a "tokonoma" (alcove). As a result, with the passage of time, the flat roof faded away and never became so powerful as to change the content of

獲得するための条件は、第1に外観、とくに陸屋根といった形が決定的だった。建物の内側では、サロンと呼ばれるような大きなヴォリュームと椅子式の床がとり入れられたにせよ、従来の日本の生活空間がそれに対決しなければならない異質なものとしては現われることが少なかった。その多くが大邸宅であったこともあるが、大きな玄関も畳の部屋も床の間も、白い角砂糖風の外観の内側で許容されてきた。だから時代とともに、この平らな屋根は消えていき、結局は日本の住宅の内容を変える力とはならなかった。
戦後の小住宅の形式を確立する努力のなかで、玄関や床の間を否定する考えは、床のあり方を問わずにおれなかった。例えば西欧風に、靴のままで生活するための床である。その異質の「床」が日本に定着しないことが分かってきたところで、「壁」がとって代った。それは生活空間に整合する架構としての壁ではない。つまり新しい住まいという統一体ではなく、伝統的な統一体のなかに、新しい部分が割り込んでくる、つまり住まいのなかに生じでくるズレを見せることが必要だった。それは、日本の建築家がつねに古い部分と新しい局面とを同時に見せる、ポレミックな姿を住宅に求めていたからだろう。

■70年後半に、伊東豊雄の「中野本町の家」に私たちは出会う。そのU字形の開いた部分を結んで循環的にしたプランは、ほぼシンメトリカルでありながら、架構の根拠をもたない、不安感ともいえる印象を人に植えつける。入口をわきにずらした円環状の白一色の室内は、軸線を失い、といってノンシャランな自由もない、流動的で静かな美しさに満ちている。3種類のスカイライト。いくつかのアルコーブ。円弧の壁。小刻みに雁行する壁。円形テーブル。天井を横切る裸の蛍光灯。伊東はこうした項目を挙げて、それを形態素(モルフェーム)と呼ぶ。建築からも生活からも切り離されたあるものの影。
この住宅はコンクリートならではの形態であり、道路からも中庭からもその長大な壁が見える。しかし家の形はどこにも見えない。高い所から見下ろしたときにはじめて、そのU字形が見えるのだが、それもプランの形であって家ではない。つまり、発掘された遺跡の大地にネガ状に掘り込まれた家の姿に他ならない。
この住宅において、それをつくったコンクリートは、単なる外壁へと廃棄されている。すなわち、外と内とを一体化してリアリティをもつ壁は、消える。その印象を裏づけるように、80年代に入ってその隣地に建てられた伊東の自邸も、家のヴォリュームを失った、いわばペントハウスだけで構成されている。
打放しコンクリートの物質の真実性を生きてきた流れは、ここで終る。しかし、そこからまた、ヴィジョンとしてのコンクリートはこれまで以上に雄弁に、かつ精密に語られはじめる。打放しコンクリートは、依然として日本の建築家にとって、建築と生活を考えていくうえでのもっとも興味深い対象となっている。安藤忠雄と東孝光と伊東豊雄の仕事は、本書に収められている。そして、彼等に結びつけられる世代の、さまざまの試みと考えを、ヴィジョンの問題として紹介するのが、この写真集の目的である。

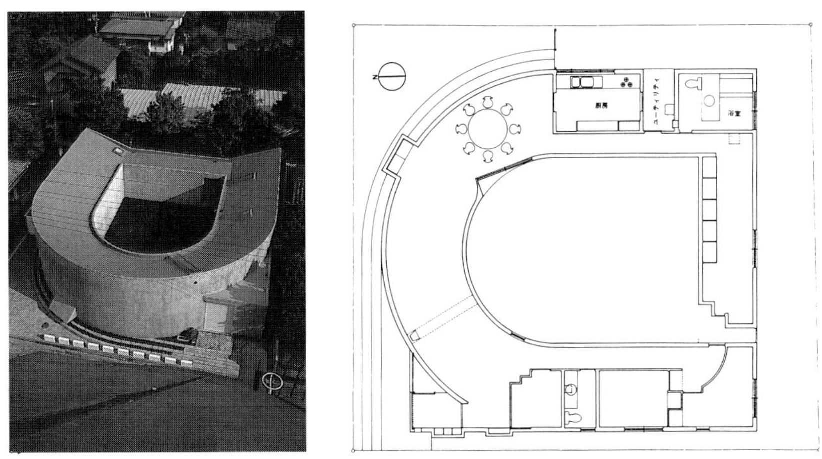
House in Nakano Honcho: ITO Toyo, 1976(photo: ITO Toyo)

Japanese homes.
Amid postwar efforts made to establish a formula for small-scale houses, the concept negating the existence of the entrance hall and "tokonoma" challenged traditional notions of what flooring should be. One Western idea, for instance, was the idea of flooring as being used with one's shoos on. And, since it soon became apparent that the heterogenous "flooring" would not take root in Japan, the "walls" took over. These, however, were not meant to serve as frameworks for separating living spaces. In other words, rather than in terms of an integrated entity in a new dwelling, it became necessary to squeeze new elements into the traditionally unified body — that is to say, to bring forth the "discrepancies" occurring in the house. And all this, no doubt, was the result of ongoing pursuit by Japanese architectonics in an effort to simultaneously exhibit the old aspects and the new ingredients in dwellings from a polemic point of view.

■In the late 1970s, Mr. Toyo Ito created his "House at Nakano Honcho." While its U-shaped plan is almost symmetrical, it gives an uneasy impression to people that it lacks a frame. Its inner room, painted all white, is a circular shape without an axis due to its shifted entrance. And, although there is no loose freedom, it is full of flexibility and quiet beauty. Citing its three kinds of skylight; several alcoves; a circular wall with another one gradually stretching out abreast; a circular table; and a naked flourescent lamp hanging transversely from the ceiling, Ito calls its style, morphology. In other words, it can be confined strictly neither to architecture or life.
This house's form seems "as if not made of concrete." A vast wall can be seen both from the road and a courtyard located at the central part of the house. However, its true shape can not be grasped from any one place. Only when looked down from above can its U-shape be recognized. This, however, is noticeable only as a plan and not as a house. In other words, it is nothing but a house of ruins discovered and imprinted on the ground as though it were negative film.
The concrete which formed this house is outcast as a simple outside wall. Ito's house, which was built nearby in 1980, is only a penthouse and therefore loses the volumetric authority of a house.
The flow out of which the physical reality of ferroconcrete architecture grew here. However, the visionary concepts of concrete continue to be articulated fluently and precisely. Architectural ferroconcrete remains as an interesting conceptual target for most Japanese architects. Works by Mr. Tadao Ando, Mr. Takamitsu Azuma and Mr. Toyo Ito are described in this book. It is our hope that this album will introduce the various concepts created by these visionaries, and some understanding of their trials, to a new generation of architectural thinkers.

Editorial Director : **AKAHIRA Kakuzo**
エディトリアル・ディレクター：赤平覚三

Design : **TAMURA Yusuke**
デザイン：田村祐介

Cooperation : **IWAMI Madoka & NISHINO Masako**
協力：石見まどか＋西野雅子

Translation : **Gendai Translation Center**
翻訳：㈱現代翻訳通訳センター

[cover project]

front cover : ANDO Tadao

back cover : TAKAMATSU Shin

Japan Houses in Ferroconcrete

1-9-12 kudankita, Chiyoda-ku, Tokyo 102, Japan
ISBN4-7661-0460-9

Printed in Japan

First printing, 1988

Works

[Explanatory Notes：凡例]

A……Atelier／アトリエ
B……Bed room／寝室
C……Court, Patio／コート、パチオ
CH……Children／子供部屋
D……Dining／食事
E……Entrance／入口
F……Family Room／家族室
G……Guest Room／客室
GE……Garage／ガレージ
H……Hall／ホール
J……Japanese Style Room／和室
K……Kitchen／台所
L……Living／居間
MA……Maisonette Apartment／メゾネット・アパート
MP……Multipurpose／多目的
S……Study／書斎
SP……Spare Room／予備室
ST……Store／ストアー
T……Terrace／テラス
V……Void Space／吹抜け

阿部 勤[ABE Tsutomu]

臼杵邸[Usuki House]

■阿部勤の住宅には、どこか馴染み深い表情がある。日本の土蔵や穀倉を思い出させる。アジアやヨーロッパの民家や倉にも似ている。石や土の壁の上に乗せられた木造の家や屋根、あるいは石や土の家に寄り添う外屋やその庇や柱といった、二重の構成が明快なのだ。
そのプランニングも単純明快である。入れ子になった2つの正方形の壁(例えば自邸)、雁行して置かれた2つの正方形の壁と、その隙間を埋める外屋的な木造部分(例えば今田邸)といった構成が、寄せ棟あるいは切妻屋根の単純な家の形にまとめられている。ところが、内部に入ると思いがけない迷路的な、現代的な生活空間に出会う。幾重にも重ねられた壁と壁との間に、多様でルーズな生活の場がとり込まれている。つまり名称のはっきりした閉じた部屋ではなく、微妙に開かれて相互に浸透し合う、どのようにも使える部屋々々の連なりがある。外観を思い切り明快な家の記号でまとめ、内側でそれを完全に裏切る、ひそかな楽しみを、建築家は狙っているようだ。
阿部の住宅の多くは、上を向いている。屋根は家をすっぽり蔽うのではなく、さりげなく開いて空の光を家の底まで入れようとしている。家の本来的な垂直性を強調したいと、彼はいう。そのために重ねられた混構造であり、その基部としてのコンクリートの壁が強く印象づけられる。

■敷地は南北に長く、西寄りに既存の住宅がある。この建物と道路に挟まれた部分に、やはり南北に長い建物を新たにはめ込んだ。東側の壁はそのまま道に面した塀にもなっている。
北端に台所とサニタリー、南端に和室をコンパクトにまとめて配し、その上部にそれぞれ寝室を乗せている。両者を結ぶスペースは、1階では居間・食堂に、2階ではサンルームになっている。同時に地下部分に、書庫、倉庫、ピアノ室が確保されている。
地下室の南端に埋められたファイア・プレイス(イングルヌック的な部屋)とその上の3帖の和室は、この家の象徴ともいえる。それは心地よい穴ぐらと風の通る方丈、冬と夏、西欧的空間と日本的空間との重層を巧妙に演出している。ここに1日籠っていれば、居間のスペースは半ば屋外的な場所に思えるだろう。また、台所・食堂上部のストア・フロントのようなガラスの壁面も魅力的だ。これは寝室階のH形の通路・ブリッジを囲う壁でもあり、さらにその上部のサンルームを温室のように蔽うガラス屋根からの光を、階下まで落とす仕掛けでもある。

■The houses designed by Mr. Tsutomu Abe contain a sort of friendly atmosphere. They summon up the air of Japanese warehouses or granneries, or the private houses and storehouses of Asia and Europe. The double constructions of house, built of stone or earthen walls, and outhouses, constructed of columns and roof on stone or earthen floors, are cleanly defined.
The designs are also down-to-earth and clean-cut. For example, within his own house one can see the two squares of the walls placed side-by-side and the external wooden building which fills the gap between. Another example is the Imada house, which is constructed to create the simple shell of the house with the use of the ridging or slanting roof. Although the appearance of the house is quite common-place, it is interesting to discover the maze-like use of the space available. Between the many layers of walls lies the various loose styles of life that may be enjoyed. In other words, each room is not distinctively separated, but all are left slightly open to create an aura of simple penetration from one room to another. They can be used singularly or together for any purpose desired.
Most of Mr. Abe's houses are built with special concentration on the upward accent. He is fond of skylights which allow natural light to filter into the rooms instead of blocking it off. By doing so it emphasizes the original vertical aspect of the construction. Owing to this, the end product seems to give off a layered feel, and this in combination with the concrete walls and foundations is impressive.

■The length of the building stretches north to south, and there is already a house existing off to the east. Another long building which also stretches north to south has been built between the road and this house. The eastern wall plays the role of a fence as it faces onto the road.
The kitchen and sanitary room have been fitted onto the northern side of the house, and the Japanese room has been compactly arranged in the southern part with the bedrooms above. The space linking these areas is taken up with the living room and dining room on the first floor and the sunroom on the second floor. In addition to this, the library, storeroom and music room have been located in the basement.
The fireplace, built onto the southern wall of the basement (a sort of inglenook room), and the three tatami-mat room above it are probably the main symbols of this house. They skillfully accent the many layers of the cosy celler and well-ventilated tiny room, as well as the western-style and Japanese-style rooms, in both summer and winter. By shutting oneself up in the living room for a day, it is easy to imagine that one is in fact outside. This effect is compounded by the glass walls in the kitchen and dining room which resemble shop windows. This wall is the major party in the construction of the H-shaped hall which is made from the staircase, which leads to the bedrooms, and the surrounding stairwell balconies. This is made even more attractive by the sunlight stabbing its way through the skylight to summon up a feeling of a sunroom or greenhouse.

1

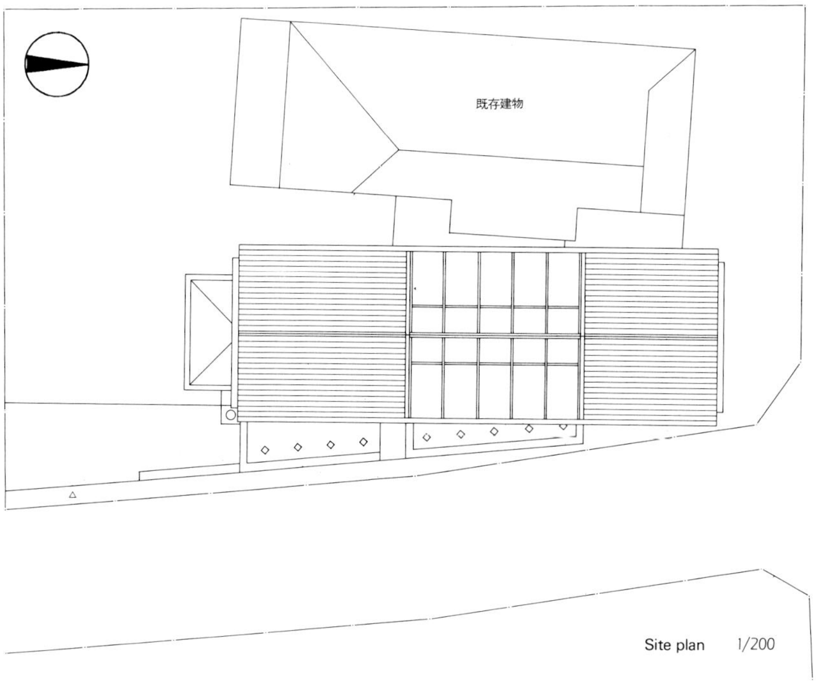

Site plan 1/200

2

3

1南東からの全景。道路に面した段状の壁は階段室。コンクリートの軀体の上に乗せられた寝室の窓と木造の切妻屋根。
2玄関のまわりはまだ未整備で、今後手を入れる予定。
3北東から見る。角地に建っており、右に見える既存の家の玄関は北側にある。

1A complete view from the southeast. The stepped wall facing the road is the stairwell. The bedroom window has been placed above the main concrete body of the house and set into the slanting roof.
2The area around the front door has yet to be worked upon, but this will be remedied in the near future.
3A view from the northeast. This has been constructed on a square site, and the front door of the existing house on the right enjoys a view of the north.

South elevation 1/200

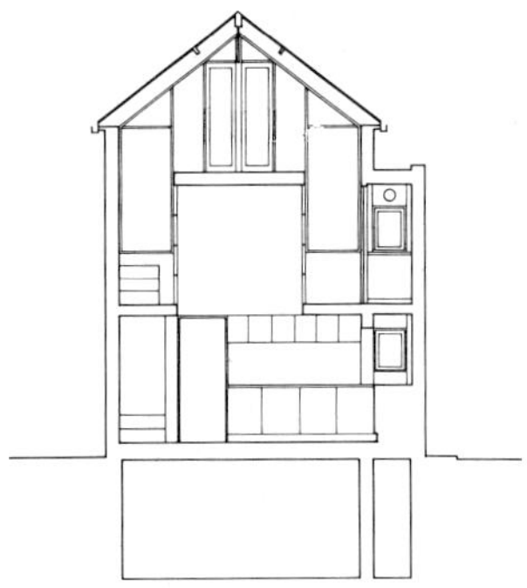

Section

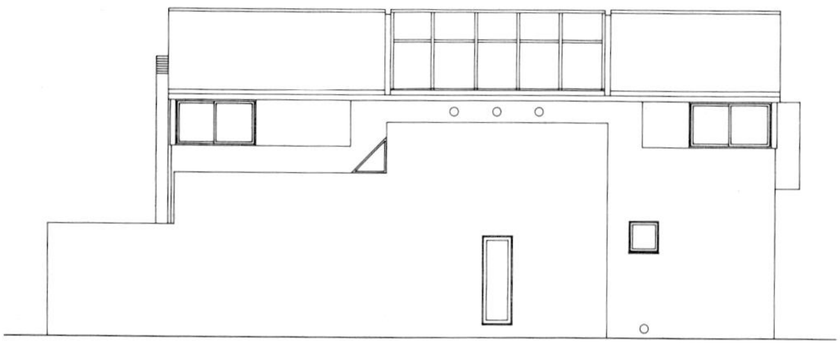

East elevation

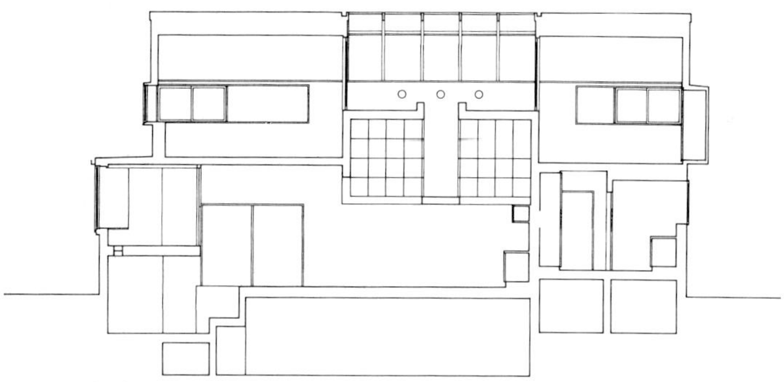

Section

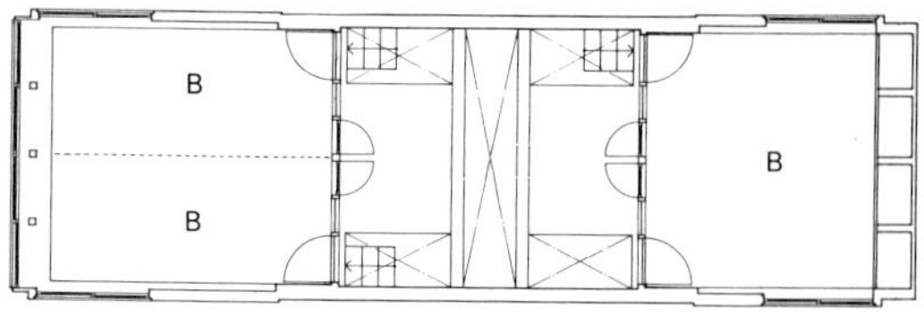

Second floor plan

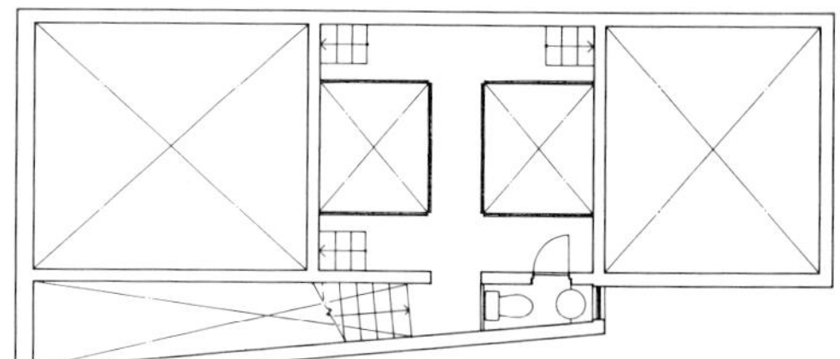

A ground plan of the mezzanine

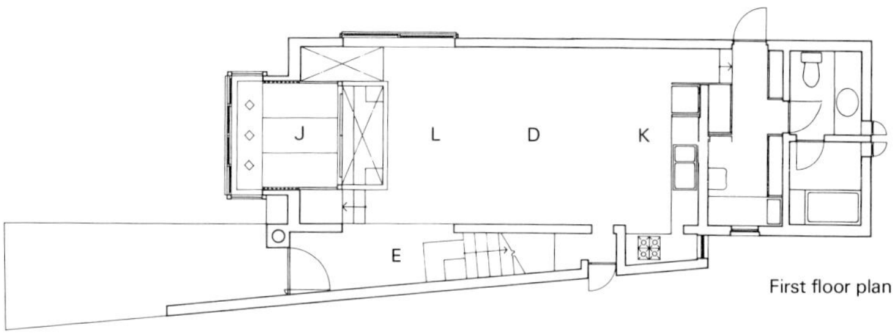

First floor plan

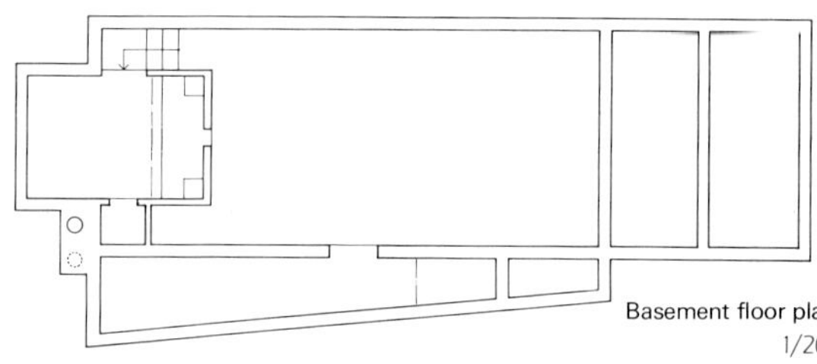

Basement floor plan
1/200

4

4 階段室から玄関ホールを見下ろす。自転車が置かれた、半ば屋外的な扱いだが、ホールはそのまま居間につながっている。

5 居間と食堂。突き当たりに台所、右手に玄関ホール。タイルの床とフローリングの床との連続。

6 食堂テーブルの真上に渡されている上階のブリッジ。ハイサイド・ライトとして扱われているガラスの壁は、部分的に開閉して風を通す。右手奥に、南端の和室。

4 Looking down upon the entrance hall from the stairwell. This hallway can be used for semioutdoor purposes, as is indicated by the presence of the bicycle, but it also leads to the living room.

5 The living room and the dining room. The kitchen is at the front, and the entrance hall is on the right. A neat continuation has been achieved from the tiled floor through.

6 A bridge above the dining table. The glass walls which incorporate the side-lights can also be partly opened for ventilation. To the rightrear a Japanese style room on the south end.

5

6

7

8

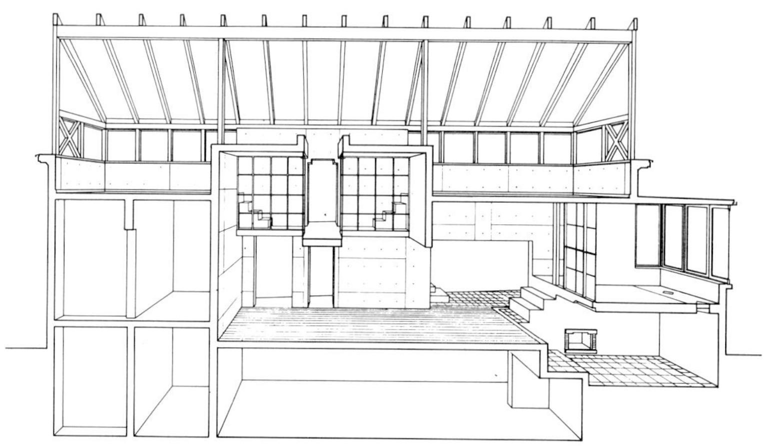

7南端の和室とその真下に埋め込まれたファイア・プレイス。食堂のハイサイド・ライトの格子の正方形パターンが、ここでも反復されている。
8ファイア・プレイスから見た、居間の床面の広がり。ここから地下室にさらに降りていく階段が隠されている。
9寝室はH形の通路・ブリッジと、その上のサンルームとの中間のレベルにある。全体として6層の床レベルで構成。

7A Japanese room at the southern end with a fireplace built right under it. The latticed-square high side-lights of the dining room have also been repeated here.
8The floor space of the living room viewed from the fireplace. The staircase that leads down to the basement is concealed.
9The bedrooms are situated in the intermediate level which is created by the H-shaped hallway, the stairwell balconies and the sunroom. In total this house has six floor levels.

9

綿貫飯田邸 [Watanuki-Iida House]

■東京郊外の住宅街に建つ3世代の家。間口が狭く奥に深い敷地であり、1階、向かって左側に薬局があるが、住宅街の中で目立ちすぎないように奥に引き込んで店のドアがあり、反対側の右手奥に住居の玄関がある。
中央部、道路に突き出ている彫刻的な階段をのぼると、広いアクセス・テラスがある。この階段と左右のアルコーブと円柱、2階のテラスを囲う壁と開口部がシンメトリカルな住宅の顔になっている。建物のファサードでありながら、実は屋上庭園のファサードであり、閉じながら開いて、生活の佇まいを垣間見せる。塀や生垣で囲った、昔ながらの日本の郊外住宅の姿。
2階まで建ち上げたコンクリートの箱の、内部の間仕切と3階屋根裏部屋は木造で白く塗装されているが、居間とそれに続く和室に、打放しのままのコンクリートの壁が顔を出し、木造部分とのコントラストをつくりだしている。
この、半ば屋外の部屋のように、東側に大きく開かれた居間は、半ば内部のようにつくられた西のテラスに心理的に結びつけられている。この両者の間にある、スカイライトの下の玄関まわりと螺旋階段が、3階の子供部屋、1階の住居、地下階を一挙につなぎ、ただの階段室以上の中心コアとなっている。いわば大黒柱だ。

■A three-generation house built in a residential area in the suburbs of Tokyo. The site with a road is deep but not wide enough. There is a chemists on the first floor to the left, and the door has been concealed at the rear so as not to stand out too much. The entrance to the house is also at the back to the right. The steps, which project out towards the road, lead up to a wide access terrace. The symmetrical facade of this house is made up of the staircase, the alcove columns on either side, the wall surrounding the second floor terrace, and the openings. These also lead the eyes up to the roof garden. Seeming open when in fact closed, it has the ability of showing "life" to those interested enough to stop and look.
The concrete box structure stretches to the second floor. The partition of the inside and the attic are made of wood and are painted in white. The concrete wall of the living room has been left untreated to create an interesting contrast with the wooden finish of the Japanese room.
This living room, which spreads along the southern side of the house, has been psycologically linked with the terrace built onto the northern side of the house. It is almost like a semi-outdoor room. The combination of the entrance and the spiral staircase, situated beneath the skylight, produces a very strong vertical centralization by linking the children's room on the third floor, the living quarters on the first floor, and then down to the basement.

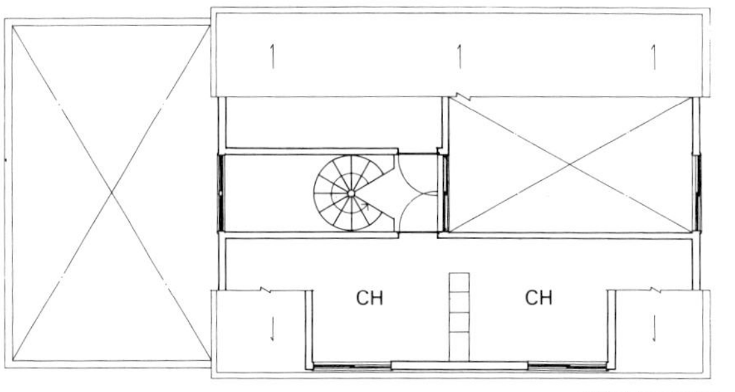

Third floor plan

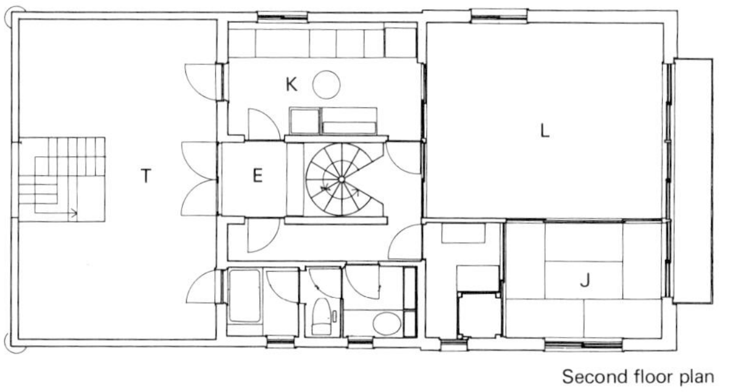

Second floor plan

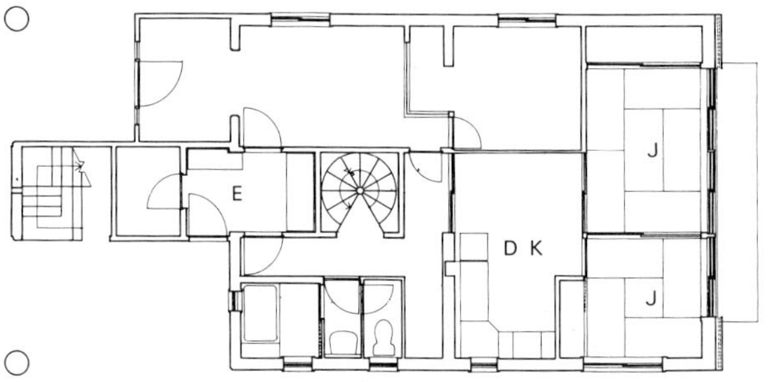

First floor plan
1/200

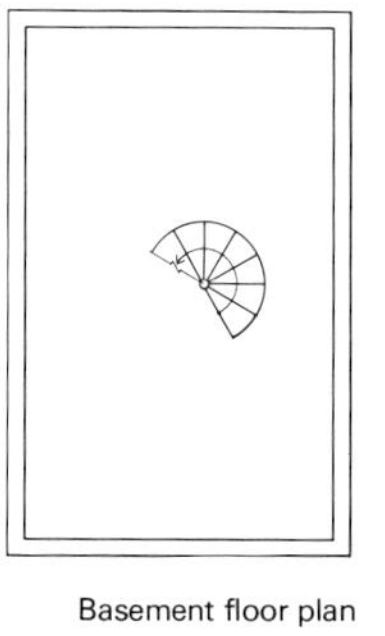
Basement floor plan

1

2

3

1道路側のファサード。左手に薬局、右手にカーポート。
2外階段をのぼると、テラス越しに切妻の家のファサードが見えてくる。観音開きの玄関ドア越しに、家の中心を占める印象的な螺旋階段が見える。台所も浴室も、テラスに直接出られるドアをもつ。

3彫刻的な階段。右手に1階住居の玄関。

1The facade on the road-side. The chemists on the left, and the car-port on the right.
2The actual facade of the house comes into view through the terrace when climbing the outside staircase. The impressive spiral staircase situated in the center of the house can been seen through the double front doors. Both the kitchen and the bathroom lead onto the terrace.

3The statuesque staircase. The front door to the house is on the right-hand side.

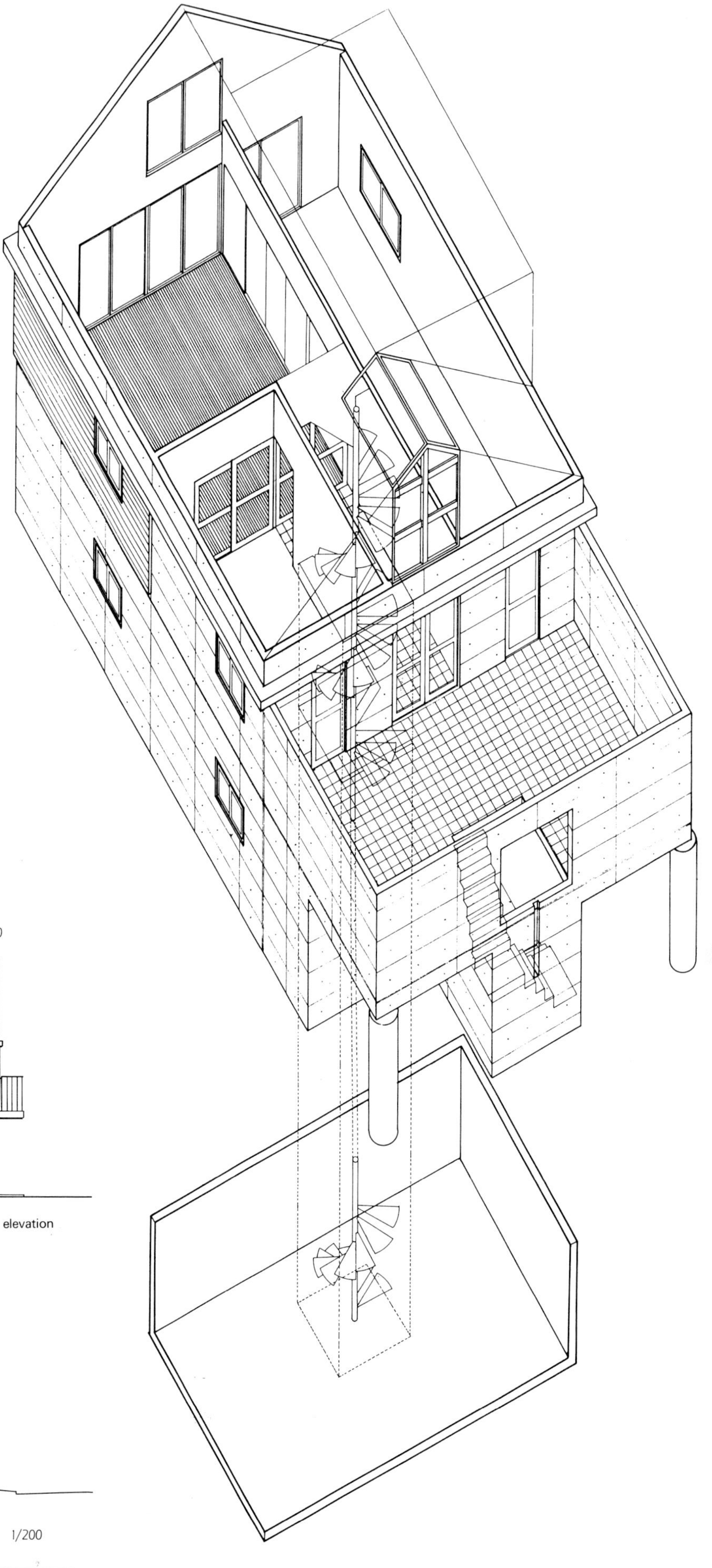

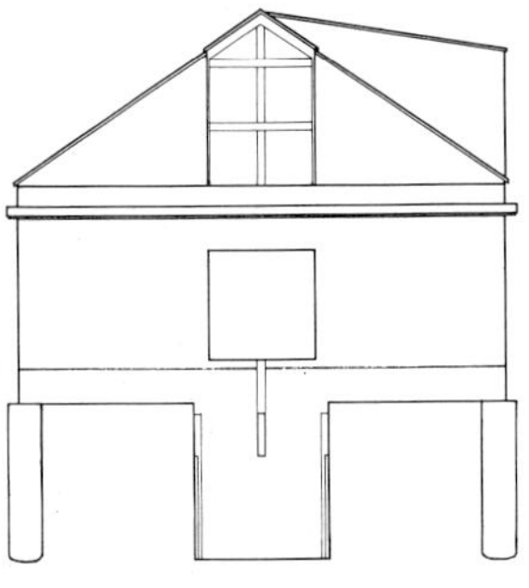
West elevation 1/200

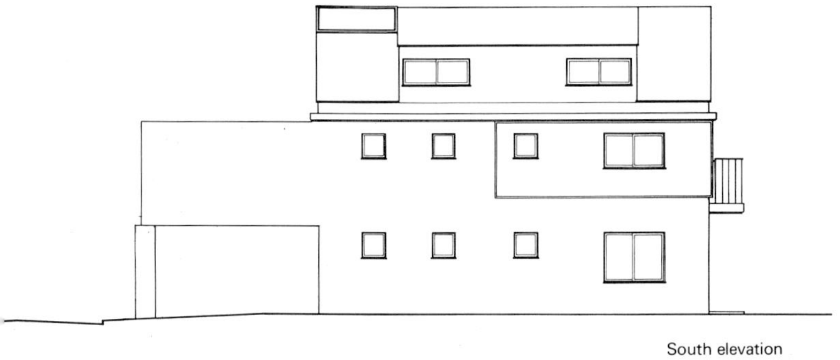
South elevation

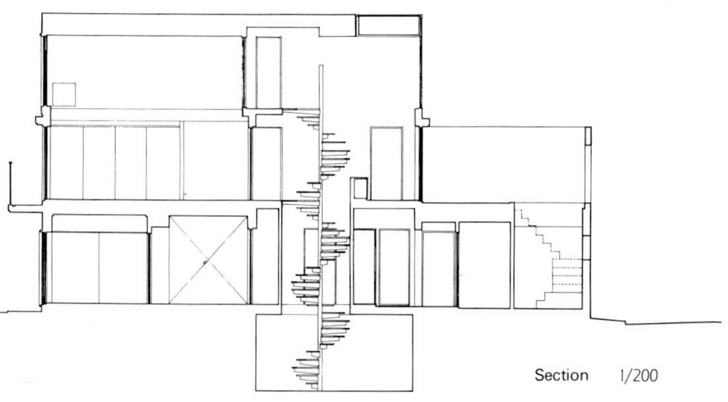
Section 1/200

4テラスから外階段を見下ろす。テラスの床は、住む人たちの手で楽しみながら仕上げられていくようだ。
5 2階玄関ホールより螺旋階段を見上げる。3階へのドアが見える。
6 7 2階。居間・和室を挟む両側の壁はコンクリート打放しのまま。この壁は外断熱になっている。

4 Looking down on the outside staircase from the terrace. The floors on the terrace have seemingly been decorated in accordance with individual taste.
5 Looking up the spiral staircase from the entrance hall on the second floor.
6 7 The second floor. The wall between the living room and the Japanese room has been left in ferroconcrete. Exterior of the wall is insulation.

5

6

7

安藤忠雄[ANDO Tadao]

細工谷の家[Town House in Saikudani]

■敷地の両隣に家が迫ってきているような都市住宅の場合、安藤忠雄は敷地いっぱいに、屋外空間をも含めたセットとしての箱を、家と家との間に挿し込む。例えば、ここに紹介する野口邸や孫邸がそうであり、他には「住吉の長屋」、「ガラスブロックの家」、「九条の長屋」等も同じである。
一方、敷地に恵まれた場合、安藤の住宅建築は、庭に半ば埋められたかのように地形を生かし、土地になじませようとする。小篠邸がその好例であり、ここに紹介する畑邸もそうである。
しかしいずれにしても、そこに登場するのは、ほとんどの場合、打放しコンクリートの直方体である。あるいはその組み合わせである。
直方体の内部はさまざまに工夫されたプランニングによって思いがけない多様で変化のある空間をつくり出しているが、直方体の輪郭が崩れることは滅多にない。1枚の壁や1本の梁が残って最後まで直方体の外形を守っている。そこには個の領域を守ろうとする住み手の意志の、直截な建築化が感じられる。その建築とは部屋のヴォリュームである以上に、何もない領域である。安藤の建築は挿入されるのではなく、抽出しのように引き抜かれたあとのヴォイトを指している。そこに、完璧な精度の打放しコンクリートが必要とされるのである。
この領域のとらえ方は、関西特有のものかもしれない。ある意味では彼の建築を見、批評する者の生活体験の背景を、逆に問い直してくるのである。

■大阪の典型的な下町、狭い道に昔からの小さな店が賑やかな顔を連ねている。その一角の棟続きの木造長屋のひとつを取り壊して、間口約3.5m、奥行き15mという細長い敷地に打放しコンクリート3階建ての家につくり直した。施主はその長屋に住んでいた家族で、夫婦と子供2人、そして両親の6人である。
全体をほぼ三等分して中央に中庭を確保しているプランニングは、かつて安藤が手がけた「住吉の長屋」と基本的に同じ構成であるが、ここでは全体がさらに縦に二分されて、各部屋のスケールは、ある限界をさらに超えて、いっそう厳密な最小限のスケールへと突入している。その結果、机や椅子、寝台の配置までが動かし難く決められ、夜昼の使い分けも必要な部屋々々となっている。しかしそれは同時に、より現実性のある生活像へと、住まいの形態を展開する契機となっているように思われる。例えば道路に面した、ガレージのような多目的の部屋（図面の表示は、居間）。ベッドだけを収めたような狭さだが、それに面しているテラスが途方もなく広い子供部屋。食堂の上のロフト風の居間兼寝室。等々、意外な使われ方が期待できる。
その最大の要因は、ただ部屋の正確な計量と配列にあるのではなく、この狭小な住宅のなかに存分にとり込んだ「屋外の部屋」にあるといっていい。

■As urban houses have no choice but to be built over looking neibouring houses, Mr. Tadao Ando uses the method of inserting blocks which contain outside space actually within the house.
For example, the houses of the Noguchi family and the Son family which are going to be introduced here — "the Sumiyoshi tenement", "the glass-block house", and "the Kujo tenement" — are good indications of this method.
On the other hand, where space in plentiful, Mr. Ando tries to bury the house within the garden in order to make the best use of the land available. The house of Ms. Koshino is a good example of this. And so is the Hata house which is going to be introduced here. Most of the houses that are introduced here have been constructed with ferroconcrete with hexahedron shapes.
The insides of hexahedron houses are designed with a surprising diversity and variety of space, but it is rare to actually change the outline of the main structure. One wall and beam has been left to maintain this shape. The architecture indicates more than just a volume of rooms, and a direct and international feel to it allows the tenant to exercise his individuality. The architecture of Mr. Ando tends to emphasize the void left behind after extraction as opposed to the act of insertion. This can be exploited to its full with the use of the accuracy of ferroconcrete. There is a chance that this style is peculiar to the Kansai region of Japan. There are cases which would make critics think again of their experience and background.

■A typical scene of down-town Osaka. Small, old shops have flourished in the narrow back-streets since ancient times. The wooden tenements alongside these shops have been demolished to make way for a three-story house in ferroconcrete finish along this area measuring approximately 3.5m across the front, and a 15m width. The party who commissioned the construction of this house is the family who previously lived in the tenement. The family consists of a man and wife, their two children and their parents.
This building has been divided into three and posesses a courtyard in the center. The planning of this construction is based along the Sumiyoshi tenement, also built by Mr. Ando. This building has been divided into two again to create many compact rooms. As a result, the layout of the desk, chairs and bed has been fixed in a certain position so each room has a distinct use. However, it also adds to the development of life-styles and gives reality to images. For instance, the multi-purpose room faces the road which resembles a garage (indicate as a living room in the illustration). And the children's room, which is so small it will only accept the insertion of a bed, faces the vast terrace. And the loft-style room on the dining room which doubles as a living room and a bedroom. And etc., many styles of utilization can be dreamed for them.
The most important element here is probably not only in the accurate measurements and layout, but the out-door rooms that have been introduced to this small and narrow house.

Site plan 1/1000

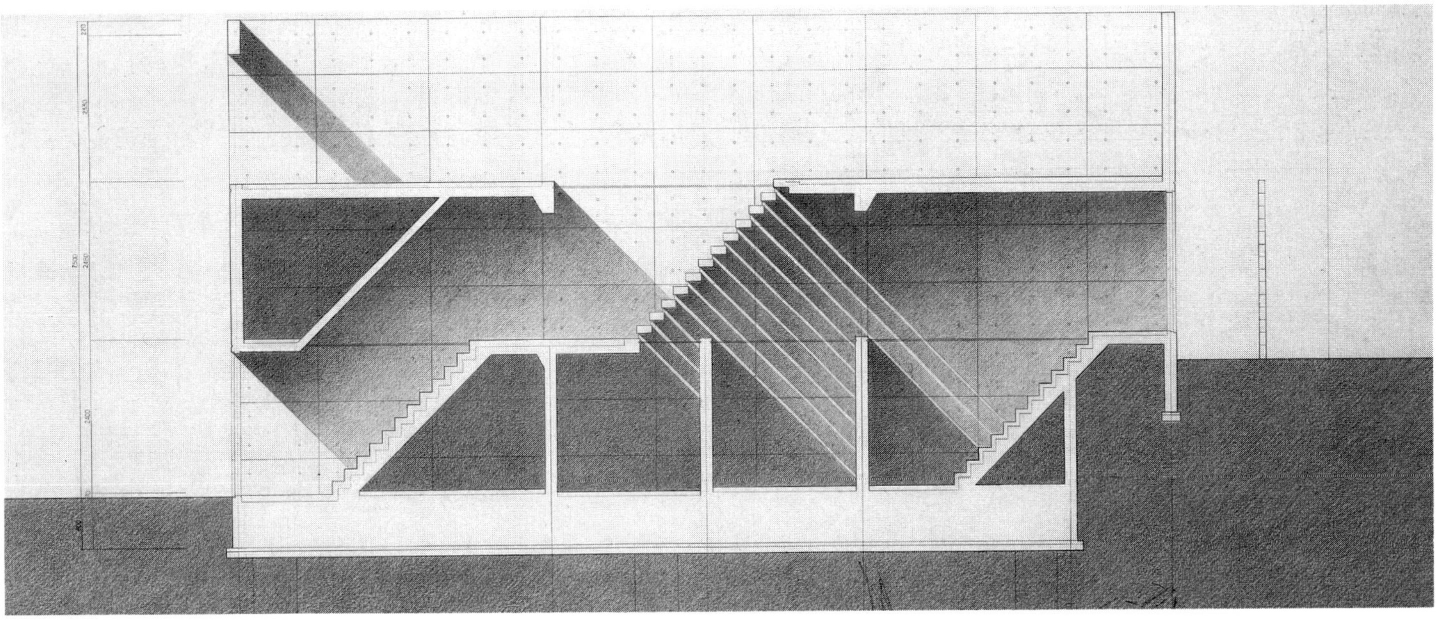

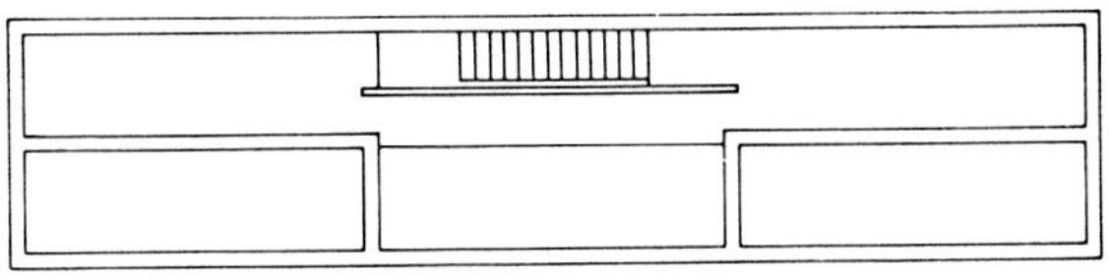

Rooftop plan

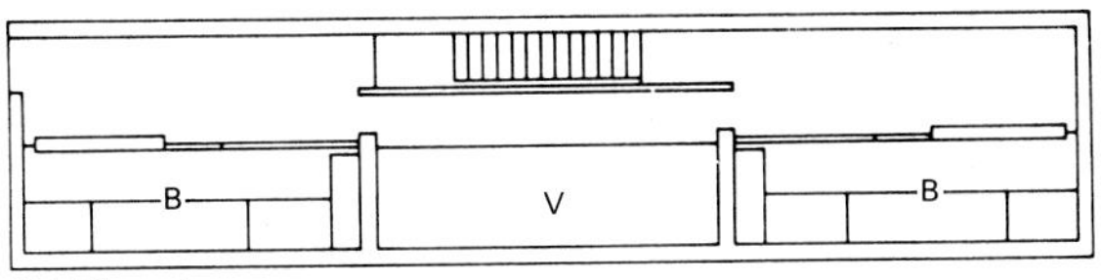

Third floor plan

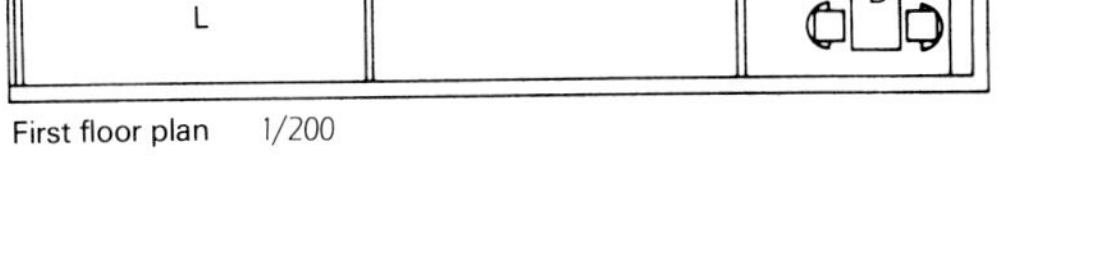

Second floor plan

First floor plan 1/200

■敷地の中央部分、全体の約1/6のスペースを占める中庭は、1階から3階まで、空に突き抜けている。3階では、奥行きいっぱいの長さを確保されたテラスが家を拡大する。

■In the center of the site, occuping approximately the one-sixth of its area, the inner garden pierces to the sky from the first to third floors. The terrace on the third floor extends as far as it can and expands the house.

2

3

4

5

6

2 棟続きの長屋の1戸を、屋根ごと切り取り、そのあとにコンクリートの箱を挿入した。
3 3階。吹き抜けと階段室との間にかかるブリッジ。突き当たりの壁と開口部は表通りに面している。
4 3階。表通り側からテラスを奥まで見通す。右手に2つの寝室が並んでいる。
5 2階。中庭上の吹き抜けに面する居間。外階段蹴上げのスリットにはめ込んだ採光用のガラス。
6 3階。テラスに面する寝室の大きなガラス。

2 One complete flat, including its roof, was removed from the apartment house and was replaced with a concrete box.
3 The third floor. A bridge connecting the well and the staircase. The wall and opening the far end faces the main street.
4 The third floor. Viewed through the terrace from the main street. To the right, there are two bedrooms in a row.
5 The second floor. Living room facing the well above the inner garden. Lighting glass fit in a riser slit on the outside steps.
6 The third floor. A large window in the bedroom facing the terrace.

7

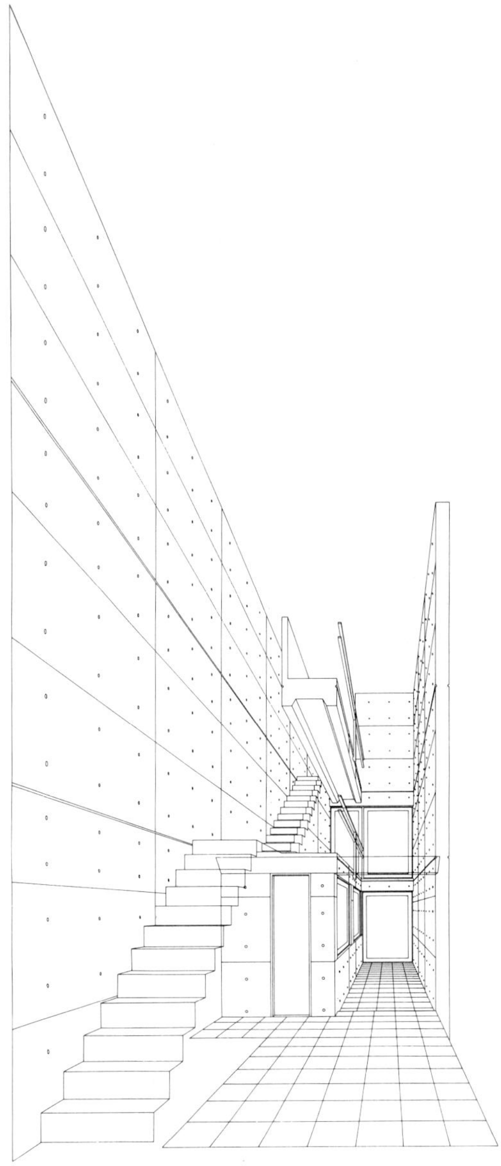

8

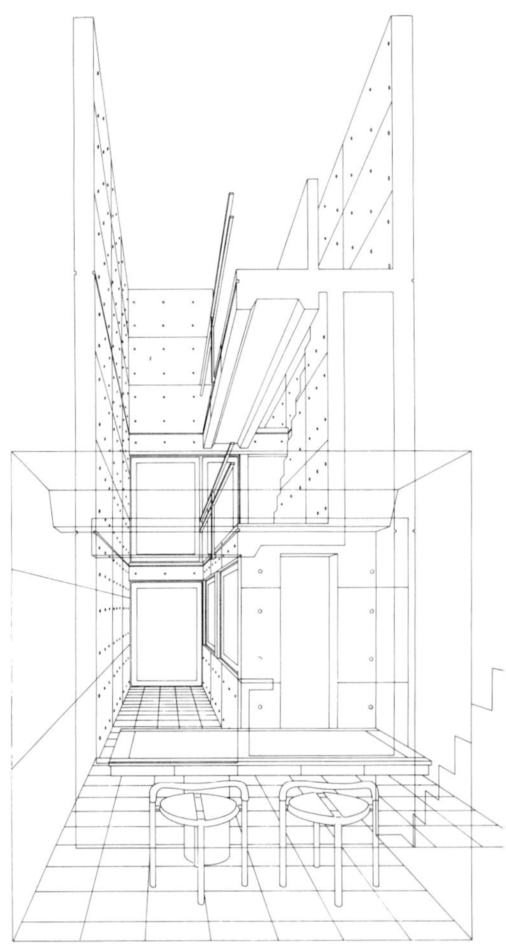

9

10

7 玄関を入って正面に、3階までの階段が真直ぐ続いている。
8 中庭を囲んで、奥の食堂、台所・浴室、道路側の居間の開口部が面している。
9 1階食堂の上部が、ロフト的に扱われた居間になっている。寝室としても使われる。
10 表通りに接する玄関わきの居間。反対側は中庭に開かれている。ガレージとも玄関ホールともいえるような佇いだが、かえって自由な使い方を誘う。
11 表通りから見る正面のファサード。左に玄関、右に居間。居間の大きなガラス戸は、通りに開け放つことができる。

7 In front of the foyer, there is a stair leading straight to the third floor.
8 Enclosing the inner garden are the dining room at the back, the kitchen, the bathroom and the living room on the street side.
9 Above the first-floor dining room is a living room, usually used as a loft. Also used as a bedroom.
10 Living room, next to the foyer, facing the main street. The opposite side faces the inner garden. It has an appearance of a garage or an entrance hall which encourages creative uses.
11 The front facade viewed from the street. To the left, the foyer. To the right, the living room. The large glass door in the living room can be opened onto the street.

11

孫邸 [Son House]

■野口邸のある細工谷の通りを、もう少し奥に進んだ同じ並びにもう一軒、この住宅が、こちらは敷地もかなり広く、しかも角地という、この町では好条件の場所に建っている。
3階分の大きなコンクリートの壁に包まれているが、通りに面したその壁の一部が2層分の高さで切り抜かれて、どこか屈託のない表情で、町のスケールやテクスチャと連続している。その開口部は巨大な窓でもあり、また、町が一軒の家のなかに混じり込む入口、あるいは住まいが町に流れ出る出口でもある。生活の内容が突然見えてくる家の切断面のようにも思われる。切断面がそのままエレヴェーションになっているのだ。
これほど開かれた印象を与えながら、玄関を入ってみると、どこまでも奥へ奥へと導かれていくような、強く閉じた方向性をこの家が隠していることに気づく。ほぼ正方形プランの中央部に、各階とも居室が集中し、その外周を階段やホールやテラスが螺旋状にとり巻いて、屋内光景や動線にある目的をもったかのような流れを形成し、居室部分に落ちついた淀みをつくり出している。2層分の高さの居間も、ここでは活気ある流れのなかに配置されていて、むしろホールと呼ぶにふさわしい。個々に切り離された部屋はなく、部屋の連なりのなかで、家全体がコンパクトに形成され、また、解かれていく。

■Going a little further along the street where the Noguchi house is situated, there is a house built on the same lines. This house is built on a large, square site that is probably the best on the whole street. Although it is confined within a large three-story high concrete wall that faces the street, this house has been sliced off at the height of two layers and maintains the scale and texture of the rest of the street and therefore does not seem out of place. The opening in the wall is a huge window; a way for the street to penetrate into the house, or a way for the house to flow into the street. It is the part of the house where the contents of life come suddenly into view. This section seems to be an elevation to the actual construction.
While giving off the impression of openness, this house conceals a strange charm of depth which is apparent upon entering the front door. All rooms have been concentrated within the center of the square building. The spiral staircase, the hall and the terrace surrounding them form a flow of direction toward the inner line of movement to create a deliberate sense of composed stagnation in the rooms. The living room, with its double-layer height, has been arranged within this current and can also be utilized as a hall. The whole house is compactly formed by the continuity of the room which have not been individually separated.

1

1表通りと路地に挟まれた角地に建つ住宅の、角の部分に入口がある。建物の本体に階段室がとり付いた構成が外観に現れている。
2通りから見て、コンクリート壁の右上部が大きく切りとられている。住宅内の眺めと外の風景とを結びつけると同時に、内部に確保された外部空間の快適さを垣間見せる。

1A dwelling built on a corner lot enclosed by a main street and an alley. An entrance to the house is at the corner. The attachment of the staircase to the house is clearly visible.
2A large part of the concrete wall at the upper right (viewed from the street) is removed. This helps to unite the views from inside the house with the scenes of the street. It also creates inside the house a feeling of pleasant outdoor space.

Site plan 1/1000

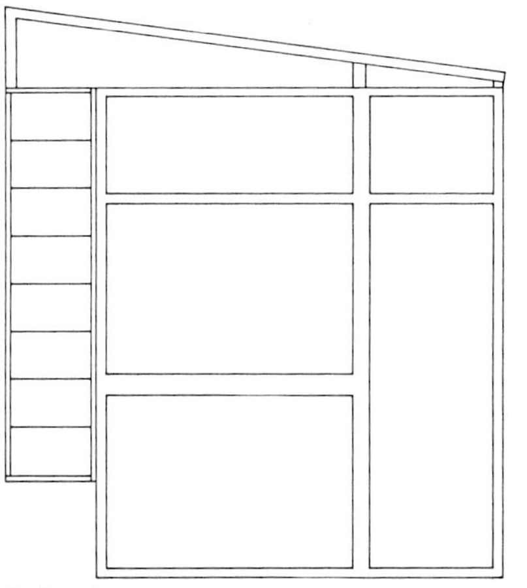

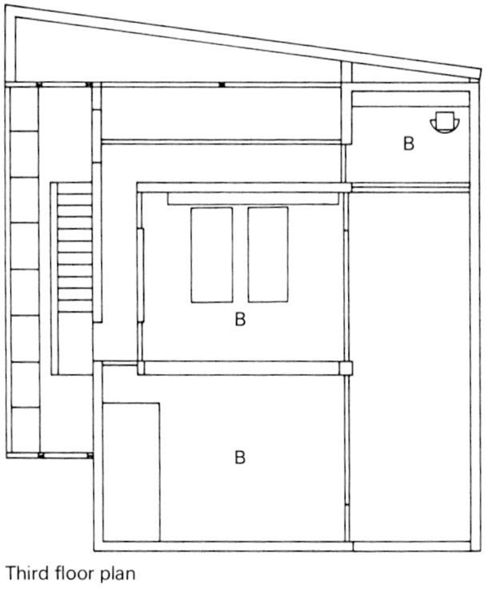

Third floor plan

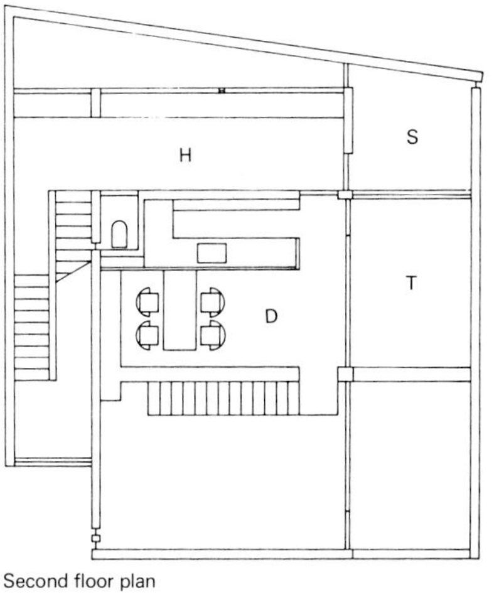

Second floor plan

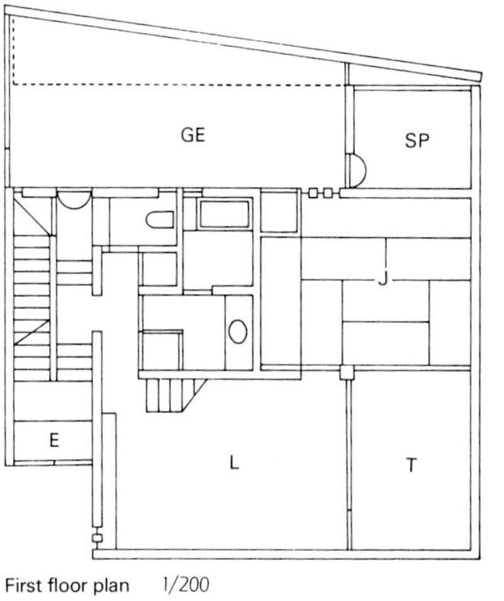

First floor plan 1/200

2

3

4

5

3玄関より階段室を見上げる。建物の中に入った途端に、各部屋へのアクセスが始まる。
4階段室。右下に玄関のドア。コンクリートの高い自立壁を挟んで階段が3階に達する。
5玄関を入った所に階段室のコンクリートの壁が迫っている。
6玄関を入り階段をのぼり、2階の家の反対側に達した所で素晴しいホールに出会う。

3 Viewed up the staircase from the foyer. Upon entering the house, we have access to each room.
4 Staircase. Front door at the bottom right. The stairs sandwiched between high concrete walls reach the third floor.
5 The wall of the stairwell draws nearer upon entering the front door.
6 When climbing up the stairs to the second floor through the foyer, there you find a wonderful hall opposite the rooms.

6

7

8

9

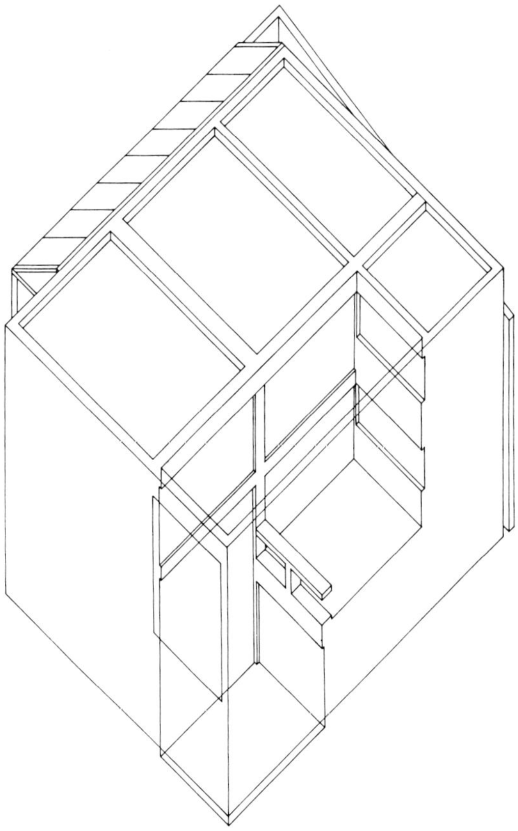

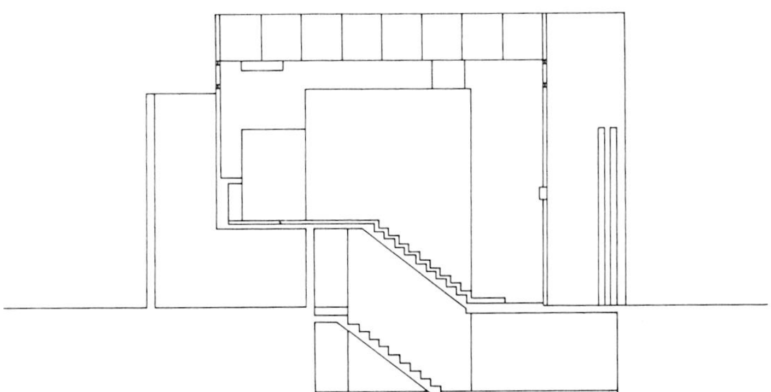

Section 1/200

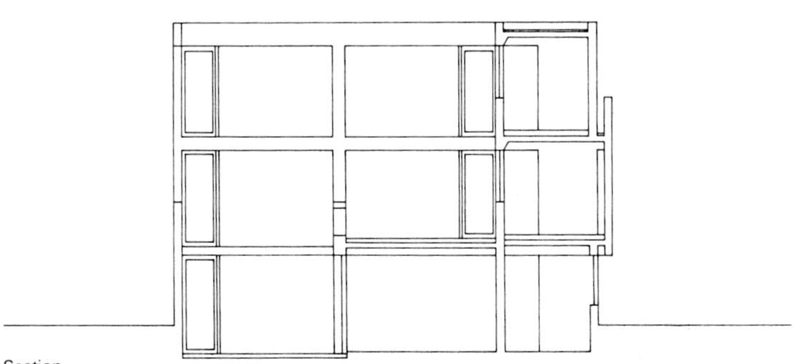

Section

10

11

789 1階居間。表通りに対しては2層分の高さの壁で強く閉じ、玄関側の壁に2本のスリット、その反対側のテラスに面して大きな開口部をとって、この部屋の向きを明確にしている。居間の中にむき出しに付けられた階段を介して、2階の食堂が一体的にある。

10 2階食堂、3階主寝室の窓が、道路側の壁を切り抜いた開口部を、緩衝帯としている。

11 食堂のコンクリートの手摺は、外部のテラスの手摺と同一線上で連続する。

789 Living room on the first floor. The side facing the main street is hidden by a two-layer high wall ; two slits in the foyer wall ; and a large opening opposite to the side of the foyer facing the terrace. These define the position of this living room. The dining room on the second floor is united by the bare-finished stairs in the living room.

10 The windows of the dining room on the second floor and the main bedroom on the third floor soften up the openings in the wall which faces out onto the road.

11 The concrete handrail in the dining room connects with the handrail of the outside terrace at the same height.

12

13

14

15

12 居間から食堂にのぼる階段。2階には1階と別のスケールや素材が待ち受けている。
13 食堂から居間を見る。鉄の手摺とコンクリートの手摺の対照。
14 15 居間の閉じた壁は、上部で食堂に開いている。食堂は居間の方向に大きな広がりをもつ。一体化した2つの部屋は、上下の床が連続する外のテラスに、さらに開かれている。

12 A stair climbing up to the dining room from the living room. On the second floor, scales and materials different from those on the first floor are used.
13 A view of the living room from the dining room. A contrast of iron and concrete handrails.
14 15 The closed wall in the living room is open to the dining room at its top. The dining room extends over the living room. The two rooms united in a body expand further to the outside terrace where two different floor levels connect.

畑邸 [Hata House]

■西宮の丘陵地に建つ。遠く海が眺望でき、敷地の前の道路ぞいに川が流れ、南に広い国有林が四季の変化を見せる。

下から坂を上がってくると、この住宅のところで道は二手に分かれ、いっそう急勾配になっていく感じがする。庭もまた、家の足元から隣地の境界まで屋根を越える高さでそそり立っている。これほど自然に恵まれた敷地のなかで、緑や土に馴染み、あるいは対立しつつ、打放しコンクリートの箱と壁が出現している。

住宅そのものは、5.7m×19.6m、3階建ての直方体にすぎない。しかし、坂の下から見上げると、全体は驚くほど多様な表情に富んでいる。建物を挟む左右の擁壁のせいでもあり、また直方体の箱に施された、単純でかつ、多方向な開閉の仕掛けによるものでもあるだろう。

敷地の中心に奥行きは深く、直方体を据えている。道路側にはガレージの開口部があるばかりで、高い壁が背を伸して立ち上がっているが、南側の庭に対しては、細長い建物の西半分あまりをさらに縦に割って、居室とテラスを段状に構成し、自然に開いている。上端に飛梁を残して直方体のイメージを損わぬよう、留意している。一方、東側には大きなボリュームの部屋を確保してその向きも転回させ、南の庭とは全く表情の違う、東のコンクリート擁壁の立体的な庭に開かせている。

自然の変化ある地形を強調するための、単純な直方体の形と素材。

■This house is located on the hill in Nishinomiya. A view of the sea presents itself in the distance, a river flows beside the road in front of the site and the change of season can be truly experienced amongst the huge national forest area to the south.

While climbing this hill it becomes apparent that the road splits into two where the house is situated, and the going becomes steeper from here on. The garden soars from the foot of the house up to the neighbouring boundaries and reaches a height which exceeds that of the roof. In an area where nature abounds, the concrete blocks and walls appear to have adapted with the greenery from one aspect, but also seems to confront it with another.

The house itself is a simple three-story, hexahedron-shaped construction with the measurement of 5.7m x 19.6m. However, despite its simplicity, a view from the bottom of the hill seems to lend it a surprisingly wealthy variety of facades. This could either be attributed to the retained aspect of the walls, or to the multi-direction openings of the hexahedron box.

The hexahedron box has been deeply placed in the center of the site. The high wall that faces the road soars from the ground with only the simple break in its bleakness being the garage door. But, in contrast, the building is wide open towards the garden and divided lengthwise to create the stepped living room and the terrace. The balance of the hexahedron image is maintained by leaving the beam at the top end. Additionally, a large room has been constructed on the eastern side of the house in order to open up the cubic garden which is contained by the concrete retaining wall on the eastern side, and which is totally separate from the southern side of the house in order to open up the cubic garden which is contained by the concrete retaining wall on the eastern side, and which is totally separate from the southern placed garden.

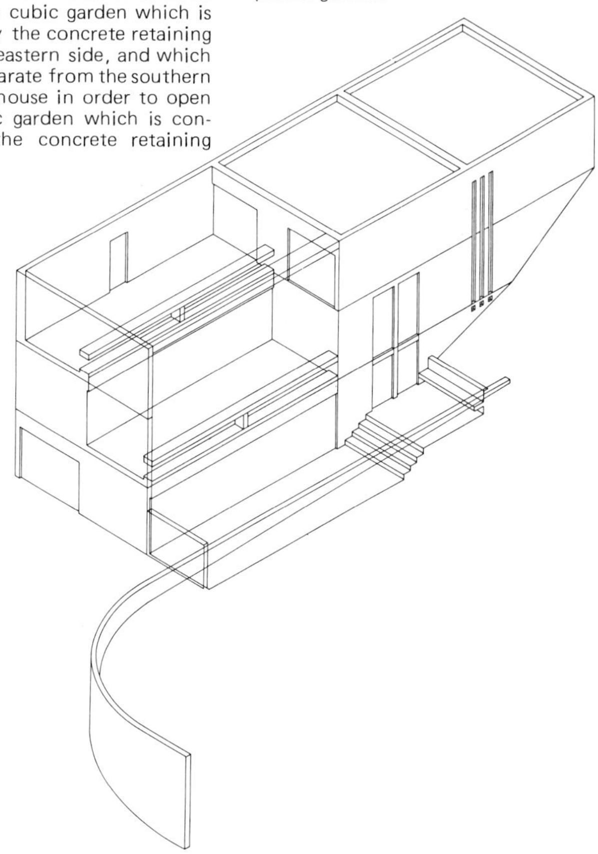

Site plan 1/600

1

2

3

4

1西側全景。急勾配の道に面して、1枚の壁、直方体、カーブする壁が並んで、家の領域をつくる。
2西側正面。左手に玄関へのアプローチ、その右へ地階のガレージと主寝室、外部テラス。
3門からゆるやかにのぼってきたところに1階の玄関があるが、直接2階のテラスに達する階段が、その傍らにある。
4建築の壁と高い擁壁に挟まれたシンプルな門。

1Overall view of the west side. One wall is a rectangular parallellepiped facing the road at a steep angle. The curved wall defines the territory of the house.
2Front view on the west side. To the left, an approach to the foyer. To its right, the garage and master bedroom underground, and the outside terrace.
3When climbing up the gentle slope from the gate, you see the foyer on the first floor. Also next to the foyer, there is a stair leading directly to the second-floor terrace.
4Gate sandwiched between the building wall and the high enclosure wall.

5

6

7

5南側正面。地下の主寝室、1階の食堂、2階のテラスが整然と重ねられている。
61階食堂とテラス。
78急斜面の庭は、底にある地階の寝室とはテラスという緩衝体を介して接し、別のところでは建物自体の壁に直接触れ合ってくる。長大な垂直の面と、波打つ水平面との出会い。

5Front view from the south side. The master bedroom underground, the dining room on the first floor, and the terrace on the second floor are piled up in an orderly manner.
6The dining room and the terrace on the first floor
78The bottom of steep garden comes up to the underground bedroom. The other part of the garden contacts directly with the building wall. An encounter between a high and large vertical face and a wavy horizontal face.

8

9

10

11

9 2階ホールから1階ホール東端の開口部を見る。東の端に開いた、2層分の高さの、この筒状のホールは、コンクリートの擁壁で趣向を変えた、垂直の庭を楽しむ部屋である。

10 1階ホールの階段踊り場から、2階ホールを見上げる。

11 2階ホールから1階ホールの南壁とスリット状の開口部を見る。

12 13 2階ホールの西に向いた開口部からは、また別の、道路側にあふれ出るかのような庭の光景と、それをすくいとっている擁壁の印象的な輪郭が見える。

9 The opening at the east end of the first-floor hall viewed from the second-floor hall. This two-story cylindrical hall opens at its east in a concrete enclosing wall with a charm of its own. The hall presents a dramatic vertical garden view.

10 Looking up the second-floor hall from the landing in the first-floor hall.

11 A view of the southern wall in the hall on the first-floor and the slit-opening from the hall on the second floor.

12 13 From the western opening in the second-floor hall, a different garden view overflowing into the street, and an impressive outline of the enclosing wall are visible.

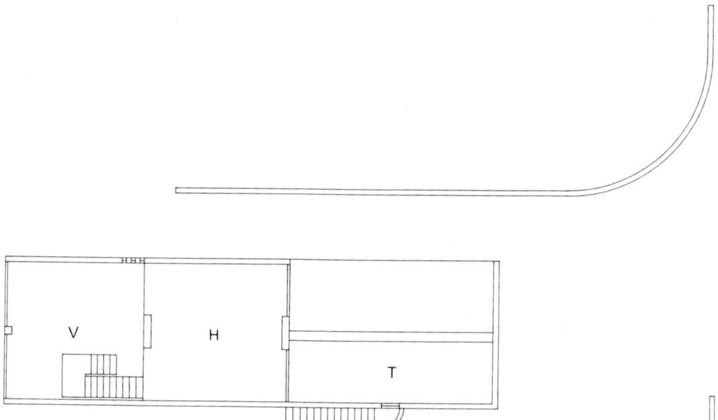

Second floor plan

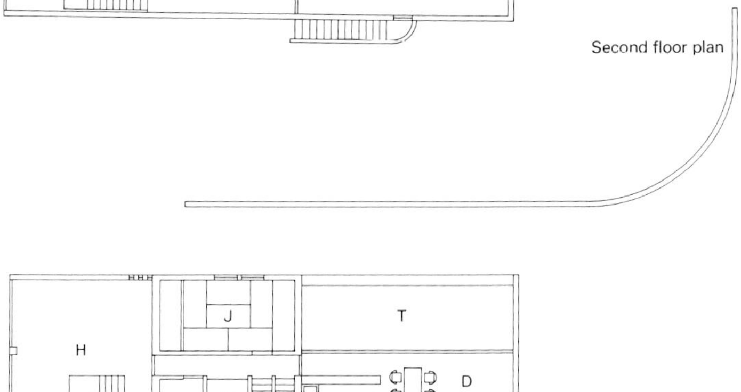

1/300
First floor plan

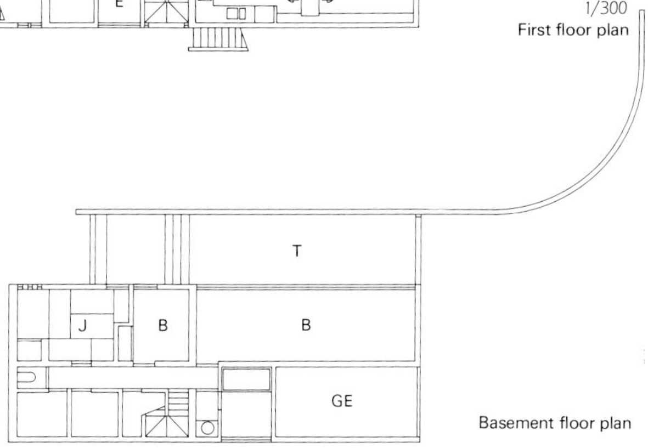

Basement floor plan

12

13

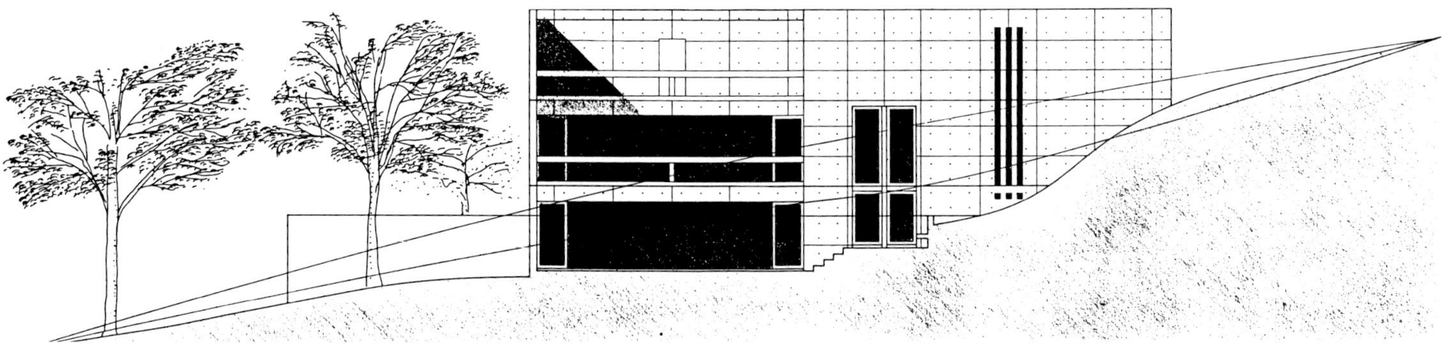

14

15

16

14 1階食堂を入口から見る。右手のテラスも、同じ形状、同じ寸法の相似の「部屋」。

15 1階ホールにむき出しにとり付けられた階段が、上下の部屋を劇的に結びつける。

16 17 玄関わきの外階段で2階テラスに直接達すると、この住宅の印象は一変する。幻想的な舞台のようなテラスを通って、2階ホールに入る。この場合、2階ホールは下の大ホールの前室ともなる。1階の玄関が住人の入口なら、2階のテラスは、招かれた客の入口として相応わしい。

17

14 A view of the dining room on the first floor from the dining room entrance. The dining room and the terrace to the right are alike in shape and dimensions.
15 The bare stairway in the first-floor hall dramatically connects the upper and lower rooms.
16 17 When climbing straight up the outside steps (beside the foyer) to the terrace on the second floor, the impression of this house changes drastically. Entering the second-floor hall through the terrace is like being on stage. The second-floor hall becomes a prelude room to the large hall on the lower floor. The foyer on the first floor is for the family; the terrace on the second floor is suitable as an entrance for the visitors invited by the family.

東 孝光 [AZUMA Takamitsu]

登美ヶ丘の家 [Tomi House]

■20年前に「塔の家」と名づけられた自邸からスタートした東孝光の100軒を越す住宅群は、コンクリート打放しのものが大部分を占め、この素材を導入した日本人の住まいの、新しい空間と生活への可能性を端的に示してきた。とくに自邸は、都心の幹線道路に面した、3.6坪6層という特異な構成に加えて内外ともに打放しコンクリートのままの、いわば垂直の洞穴のようなスペースであり、ここに20年間住んできた自信が、彼の手掛けた数多くの住宅に影響を及ぼしている。
東に先行する世代の建築家たちが固執していた平面計画の整合性、あるいは構造的なダイナミズムは、東のコンクリートの住宅群においては、より柔軟な、家の内外の境界を曖昧にしていくような間取りやディテール、そして地中海の民家集落を思い出させるような、素朴で多様な壁の表情にとって代わられている。
ここ7、8年、東はさらに新しい試みを行っている。敷地全体を立体的にパッケージするようなコンクリートのマッスをまずつくり、それをいくつかに分断し、あるいは内側を大きくシリンダー状にくりぬいて、一軒の家を何軒かの家の集合に見せるような手法である。統一的な集落とアナーキーな家の集合体との間を、彼はまだ往復している。

■郊外住宅地の、勾配のある角地に建つ。敷地全体をひとつのボリュームと見なして、そこから家と庭を彫刻的に切り出している。道路面にカーポートとガレージ、門を配し、家そのものと庭をその上部にあげている。
玄関までの階段と段状の壁、コンクリートの擁壁にせき止められたような庭の立体的な扱いに対して、南側の中央に立つ円柱を軸として内外に拡がるコンクリートの円環状の梁が、水平的な家の強い表情となっている。
この「巨人の指輪」とでもいうべきエレメントは、東の80年代に入ってからの住宅に繰り返し登場している。それは家の中心を形成するというよりは、家の全体を表している。自由に配された、いわば遠心力をもって外へ延びようとするそれぞれの部屋を連結し、家の形に収まることなく輻輳した生活の場の、全体の姿を見せようと苦心している。
主寝室、子供部屋、和室は引き戸ひとつで開閉し、円弧状のギャラリーや食堂と融合し、その内側にとられたアルコーブ的な居間へ、さらにテラスへ、庭へと段階的に連続していく。南側の露出した円弧の梁を基点として、将来は増築される可能性も考えられている。居間は、より微妙な内/外空間となるだろう。

■Mr. Takamitsu Azuma built his own house, called "Tower House" twenty-years ago, and since then has constructed more than one hundred. Most of his constructions have been built with ferroconcrete to summon up the indication of the potential of new space and life than is indigent to Japanese houses. His own house is in fact a volume of space built like a vertical cave. It is a unique construction of the ferroconcrete that stands six-stories high but has its foundations in only 12m² of land. It faces onto a highway that leads into the city, and the confident determination which the building exudes has been a perfect influence on the many houses that he has built over the last twenty-years.
The co-ordination of ground planning and constructive dynamism that was created in the generation of architects preceeding Mr. Azuma have been replaced with a more flexible style of planning which seems to feed on a kind of uncertainty when it comes to boundaries and details, such as the huge variety of simple wall facades that one would expect to see in the Mediterranean.
Mr. Azuma has worked on a new concept for the last seven or eight years. It is the method of making one house seem to be a gathering of several houses by separating and coring the mass of concrete that is prevalent around the site. An attempt to achieve urbanity within one residence; a house as a colony.

■This house is located on a piece of square land on the slopes of residentail suburbia. Giving consideration to the complete site in the sense of volume, this house has been created in the same way as a statue might have been. The actual residence and the garden have been constructed on top of the garage. The garage and car-port and the gate face onto the road. The circular line of concrete, which spreads the inside and outside with the use of a column erected to the south, acts as an axis upon which the house rotates and is in direct contrast to the vertical strength summoned up by the main wall, the staircase and its serrated wall which leads to the entrance, and the cubic garden which appears almost dammed up by the concrete retaining wall.
This style of cement, known as "the Giant's Ring", has appeared repeatedly in Azuma houses since 1980. It manages to express the feeling of the whole house as opposed to simply forming the center of the house. Each room has been arranged with a type of freedom that stretches towards the outdoors upon centrifugal forces and exudes the impression of sophisticated living quarters.
The main bedroom, the children's bedroom and the Japanese room can be opened and closed with a single sliding door which leads to the arc-shaped gallery, the dining room, and even then continues onto the alcove-style living room, the terrace and the garden in order. A plan to build an extention onto the exposed arc in the southern part of the house is already under way. The living room will become a delicate half-way house between the great outdoors and the house.

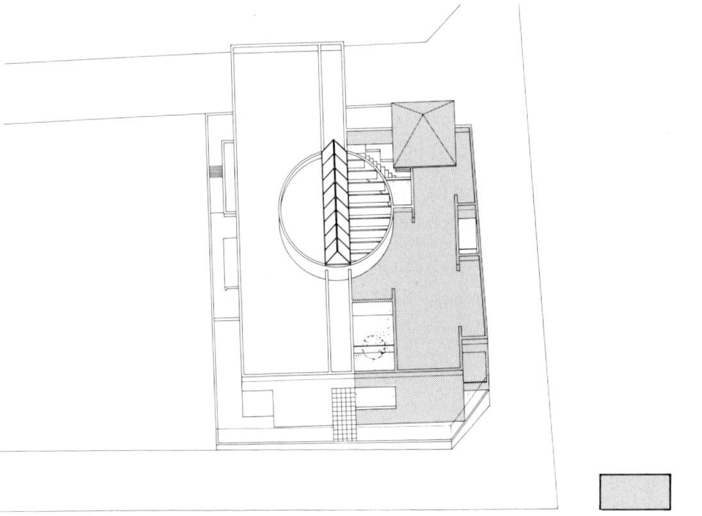

A part aimed for the second renovation

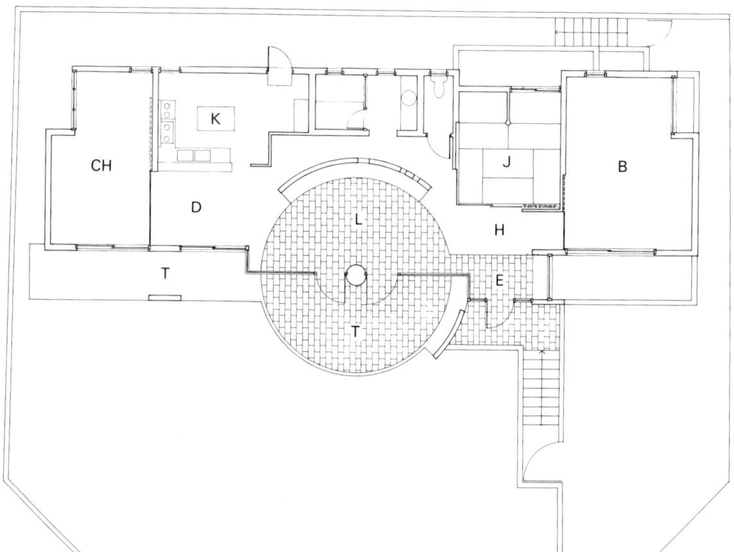

First floor plan 1/250

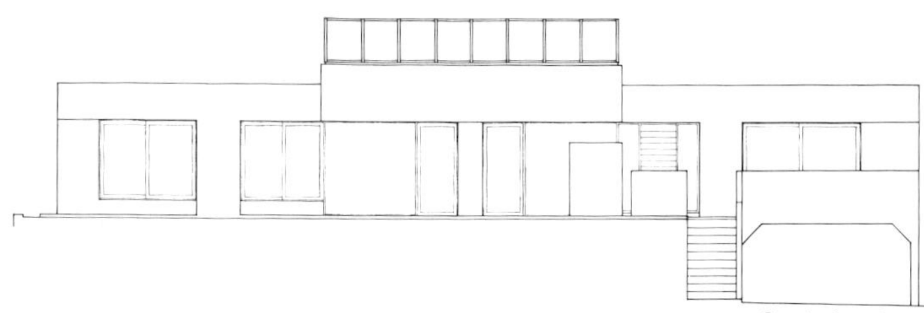

South elevation

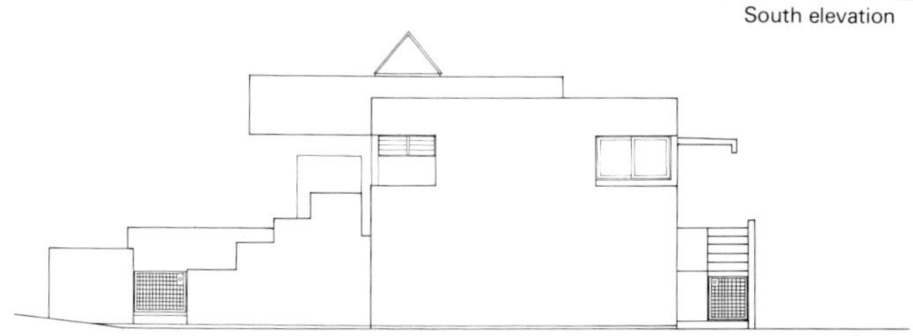

East elevation 1/250

❶南東からの全景。円環状の梁の上に乗せられたスカイライトは、増築の完成したときに、より効果的になるだろう。
❷すでに半ば内部化したテラスの表情。

❶A complete view from the southeast. The skylight on the circle will be effective when the extention is completed.
❷The terrace which has become a semi-indoor area.

1

2

3

4

5

36居間の床はそのままテラスに連続している。居間の円弧状のベンチは、そのまま食堂の床に連続している。食堂の奥に子供部屋。子供部屋は間仕切で自由に調節できるように、広い一室空間のまま。
4居間から庭を見る。増築されたあとは、この「大黒柱」やタイル貼りの円形の床の表情も一変するだろう。今はあまりにも大きく見える梁も適度なスケールになるだろう。
57円弧の壁の裏側、通路に突き出した感じの和室。障子の開閉で全く趣きが変わってしまう。

36The floor of the living room continues out onto the terrace, and the circular arc-shaped bench in the living room stretches through to the floor of the dining room. A children's room is located at the back of the dining room. It is purposely large to enable it to be divided with the use of screens.
4A view of the garden from the living room. The central pillar and the tiled floor will change when the extention has been completed. Even the beam that stands out so clearly will shrink into insignificance.
57The rear side of the arc-shaped wall. The Japanese room projecting into the road. The whole appearance can be altered by the simple act of opening or closing the paper door.

6

7

遊ハウス[YU House]

■早稲田の住宅街の細い路地に面した、70㎡近くの敷地に建つ。岩をくりぬいて棲み処としたような、コンクリートの壁が身近に迫り、またその大きな断面が最上階までもちこされている、重厚な壁の家。
映画監督とダンスをやられる夫妻2人暮らし。限られた面積のなかで、2階に書斎、地下に踊りの稽古場が確保された。後者は応接・居間を兼ね、食堂とアイランド・キッチンのスペースは最小限に抑えられている。
むしろ、玄関ホール、通路、階段が、それぞれの部屋と応答しあう第2の居住空間として、全体が立体的に屈曲しながら延びる一室空間として感じられる。東の自邸に近いといえる。斜め梁や階段の腰壁を鋸のようにギザギザにしたり、床もフローリングやタイルや玄晶石など硬質の素材で仕上げるなど、最小限ともいえる内部空間をより凝縮し、洞穴のように見せているデザインが、かえって親密な効果をあげている。書斎と階段室の境の壁に開けられた円形の穴や、台所と脱衣室の間に切り込むようにとられた坪庭も、それゆえに強い光をとり入れている。
こうした重厚さを増幅するように、家形のファサードが二重三重の独立壁として反復され、コンクリートの素材感を受け継いだ角材の格子が、坪庭や門の効果的なアクセントになっている。

■This house is constructed in 70m² of land facing a narrow alley in the residential area of Waseda. This house has ponderous concrete walls. They chich gives of the impression of being hewn from rock, rise near to us and their large open sections extend to top of this house. The study and dance studio have been constructed within limited space on the second floor and basement respectively. The latter is combined into the reception and the living room, and the space allocated the island kitchen and dining room has been kept as minimal as possible.
The entrance hall, corridor and staircase are a little too large to be considered for anything other than their purpose, but they respond nicely with the adjoining rooms and seem to wend their way through the house in an almost cubic fashion. It is obviously a Azuma house. The limited space within has been compressed even further by the use of diagonal beams, zigzag stair panels and unyeilding materials, such as tiles and stone, for the flooring. However, the overall effect is a bit like that of an intimate and comfortable cave. The circular hole has been inserted between the study and stairwell and leads to the indoor garden which lies between the kitchen and dressing room. It is almost as if it has been gouged out of the concrete in order to let in as much light as possible.
The facade of the house has been layered two and sometimes three times on independent walls to increase the ponderosity of the structure. The square timber lattice-work creates a wonderful accent for the indoor garden and gate, and seems to inherit the texture of the concrete.

1

12 狭い露地のなかできわめて印象的なファサード。家形の独立壁が、家と道との間のクッションをつくり出している。角材で組んだ屋根の形や、コンクリートの壁に一体的に鋳込んだような「家紋」は、この住宅の重厚さと楽しさを端的に見せている。

3 曲線を描く壁と屋根形の壁、打放しコンクリートと彩色された角材との対照。

12 A most impressive facade in this small alley. The independently layered wall creates a cushion between the house and the road. The shape of the wall and the family crest inset into the wall show the frank pomposity and fun contained in this construction.
3 The contradiction between the curved and roof-shaped wall, and the ferroconcrete and square timbers.

2

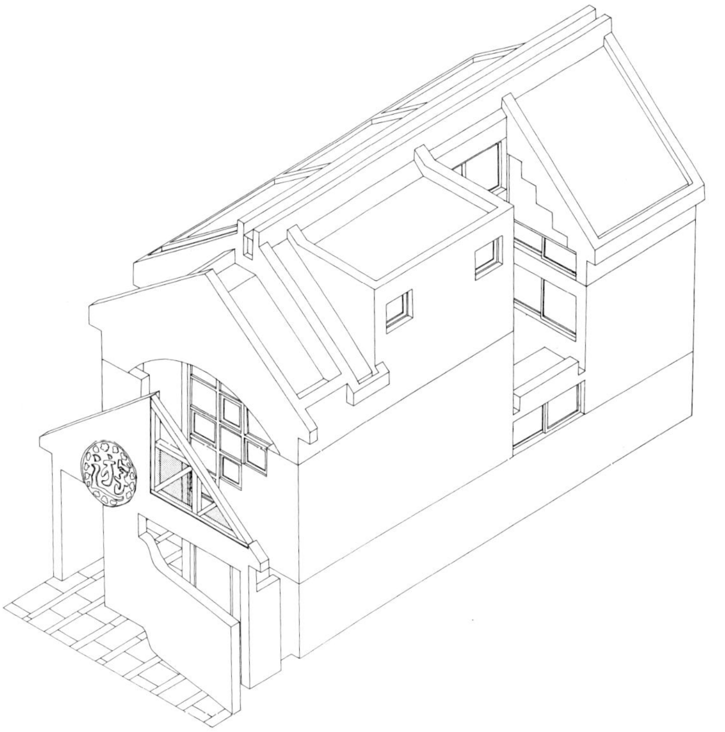

3

4

5

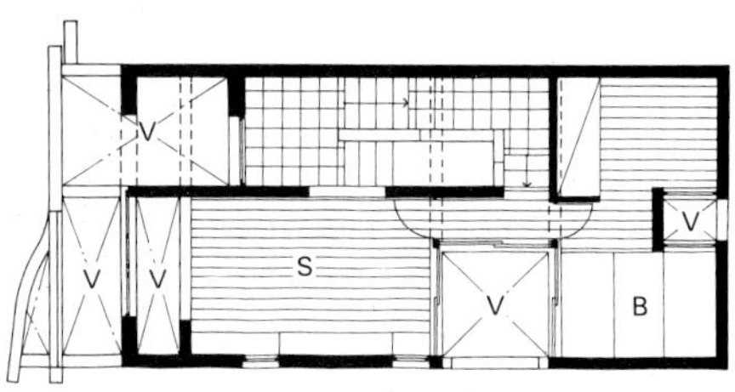

Second floor plan

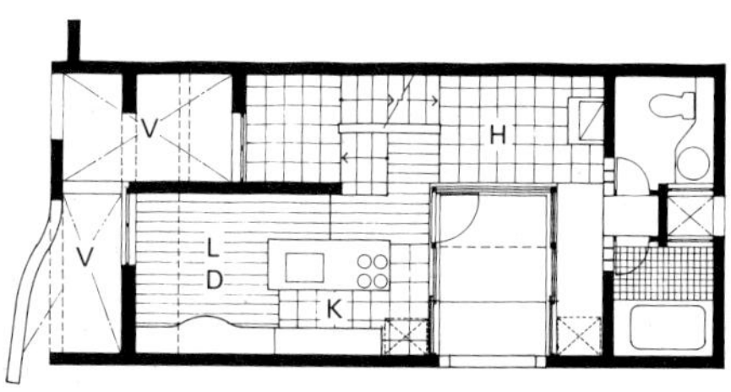

First floor plan

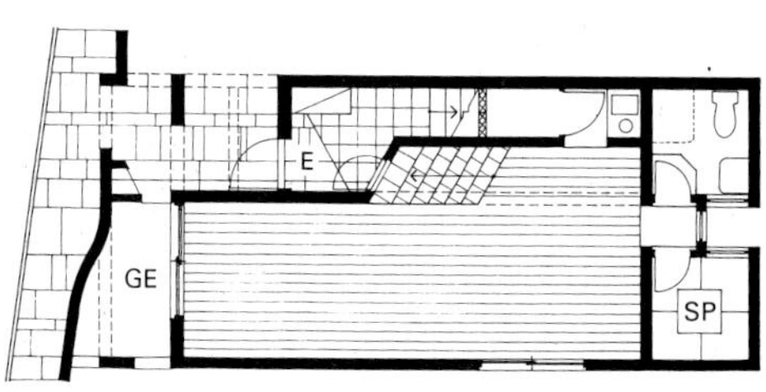

1/200
Basement floor plan

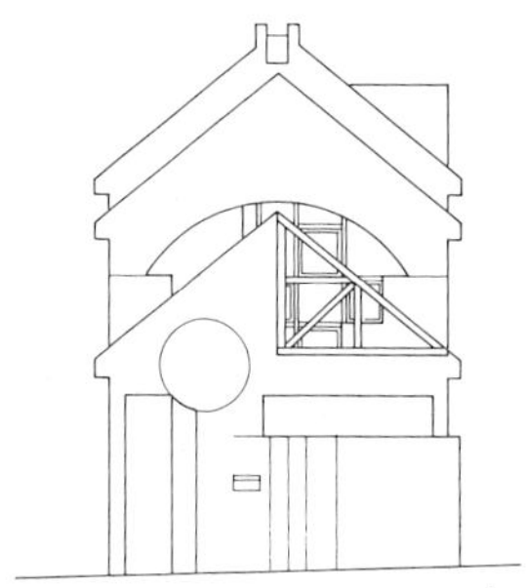

West elevation

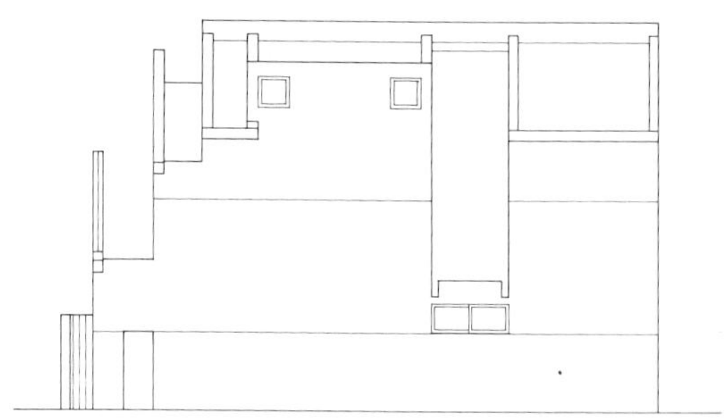

South elevation 1/200

4 切妻形とアーチ、大きな装飾がすべて打放しコンクリートで表情を少しずつ変えて見えてくる。ここには住む人と建築家との対話も感じられる。
5 玄関前のポーチ。狭いながら三重の壁・柱・梁が、この場所を魅力的にする。
6 入口から地下の稽古場に降りる階段。

4 The slant and arch. The large decoration is capable of being altered with the use of ferro-concrete. One can almost sense the discussions that took place between the owner and the architect.
5 The front-door porch. Although it is narrow, the layered wall and pillars give of a nice feeling of charm.
6 The staircase leading to the dance studio in the basement.

6

7

8

9

10

7 1階ホールより。左奥に台所・食堂。
8階段室。左下に1階の台所・食堂の床面が見える。円形の開口部の内側は書斎。屋根勾配に沿って窓の形が決められている。
9坪庭。2階では書斎(写真手前)と寝室が「外部」を挟んで向き合う。角材はオイルステインの赤。
10身体に迫ってくる三重のコンクリートの壁。しかし手前の階段はそのまま食堂に開かれて、ベンチのような風情。

7 A view from the hall on the first floor. The kitchen and the dining room spread out at the rear.
8 The stairwell. The floor of the kitchen and dining room can be seen on the lower left. Inside of the circular opening is the study. The shape of the window is in harmony with the slope of the roof.
9 The indoor garden. The study (in the foreground) faces the bedroom across the garden. The square timbers have been painted oil-stained red.
10 The three-layers of the concrete wall seem to draw near. The stairs on the near-side open into the dining room, and have the appearance of a bench.

11

12

11 規格寸法そのままの角材はオイルステインで赤や黄に塗られている。コンクリートを斜めに打って黄色で直接ペイントしたところは、ポスト。
12 道路から入口まわりを見上げる。
13 通路から台所・食堂を見る。左に坪庭。
14 左が書斎、右手に階段室。間を仕切るコンクリートの壁。スカイライト、床に接した窓、円形の開口部の適確な処理。
15 階段室。正面下に1階ホールと坪庭。

11 The regular-size timbers have been painted in red and yellow oil-stain colors. The concrete post has also been painted in yellow.
12 Looking up at the entrance from the surrounding road.
13 A view of the kitchen and dining room. The small garden leads off to the left.
14 The study is on the left, and the stairwell is on the right, with a concrete wall dividing them. The skylight and floor level window, and the circular opening.
15 The stairwell. The hall on the first floor with the indoor garden spreading out beyond.

13

14

15

原 広司[HARA Hiroshi]

夢舞台[The Stage of Dreams]

■最初期の原広司の住宅、例えば約20年前の伊藤邸を見ると、彼独特の建築の科学、例えば開口部の機能と大きさと位置の厳密な設定によってつくられた木造住宅が、伝統的な日本の木造住宅に程遠い、もみくしゃにされたような形態に、驚かされる。不定形として現れる厳密さ。
次には、傾斜地に半ば埋められた長いコンクリートのチューブ状の粟津邸が発展して、もう一度、素朴な切妻屋根の木造住宅が現れる。(自邸およびそれ以降の線対称住宅)しかし、その内部には、「都市を埋蔵する」ために部屋というよりはいくつもの家が並べられるという、従来の建築計画から逸脱したものに、しらぬ間に、住宅は変貌している。
その次に、現在に至るまで彼が次々と試みている、「多層化」による住宅群を見ることができる。彼の建築は「雲母のように」重なり合う、透明、半透明な層によって出現する。例えば、ガラス、表面を傷つけられたガラス、スリガラスなどから成り立つような家。そのなかではコンクリートも透明化へと、同時に透明性を力強く可視化するものへと向かう。
原の住宅には、住宅の既存の形式やスタイルがそのまま見えながら、意味を裏返されたりズラされたりして、慣習的な了解からもっとも遠いところに現れる、間取りや外観や素材が満ちている。

■静岡県の伊豆に建つ別荘。地震や台風に見舞われることの多い土地であるために、コンクリート造を採用したという。整然とした打放しコンクリートの骨組に、木、ステンレス板、石綿スレート板、ガラスなどが、一見不規則な形で絡みついている。
道路側のファサードは波打って、そのまま室内のホールや食堂や子供部屋の波打つ壁面に伝えられているが、見晴らしのよい北面のテラスに近づくにつれて、開口部は大きく透明になり、ぴんと張られたスキンになっていく。それは空気を失ってペチャンコになった気球が、ヘリウムを吹き込まれて端のほうから次第に膨れていく様子にも似ている。眺望のよさが建築のなかに、温度分布図形のような曲線を描きながら、浸透している。
施主が名づけたという「夢舞台」の名称は、この建物全体が、本当の住宅ではなく、演劇的装置であるようにしたいという建築家の意図に対応している。つまり閉じられたヴォリュームとしての部屋はなく、幾重にも重ねられた、そして重ねられていることを印象づけるように工夫された皮膜があり、その隙間に部屋としての機能が働いている。この「多層構造」は、原のその後の仕事でさらに展開する。

■A study of the Ito house that was built by Mr. Hiroshi Hara over twenty years ago will display his unique architectural style which uses the strictest consideration of available space and function. The size and positioning of the house is far from the traditional style of Japanese housing. It is almost as if the structure has been knocked about to produce a sense of strict discipline within a undetermined shape.
The development of the long tube-like concrete Awazu house which is half buried into the slope came next, and once again the simple wooden structure with sloping roof is apparent. His own house and others since have been built along linear symmetrical lines, but the inside holds the theme of "Bury the City" instead of just rooms. This in fact means that the design of the houses have departed from the former architectural styles of design and have been transfigured into new techniques without any prior warning.
And, there are the multi-layer structures that have also been holding their own. Mr. Hara's style of architecture tends to consist of transparent or semitransparent layers, such as mica. For example, there is a house that was made completely of clear, scratched and opaque glass. It is possible to make concrete seem transparent, and simultaneously enforce the strength of the transparency.
Although the patterns and styles of Mr. Hara's houses can clearly be seen, in fact his architecture is a lot deeper than it seems. The planning of the structure, the appearance and the materials used are always widely varied.

■This villa is located in Izu, Shizuoka prefecture. It is thought that concrete was used as a safety precaution against the many earthquakes and typhoons that prevail in the area. The wood, stainless steel sheeting, slate and glass have been entwined around the structure into irregular shapes.
The facade on the road-side tends to send waves crashing against the walls of the hall, the dining room and the children's room which lay within. However, when one approaches the northern side of the terrace with its commanding view, it seems to take on the effect of a tight and transparent skin. It is reminiscent of a flat air-ship gradually being inflated from one end by the helium gas. The advantage of the fine view has been compounded by the curve which is like a temperature chart graph. The name of the house, "the Stage of Dreams", was chosen by the owner with the intention of summing up the overall feel of the architecture as a piece of dramatic equipment as opposed to an ordinary house. In other words, instead of allocating a certain amount of volume to each room and dividing them accordingly, each room is separated from each other by an impressive amount of filmy membranes. This multi-layered style of construction appears again in the later work of Hara.

1

1 道路側から見る。「多層構造」の手法は塀にまで及んでいる。
2 反対側(北側)からの見下し。コンクリートの骨組に対して、ガラス面が斜めにズレている。
3 道路側ファサードの詳細。

1 A view from the road. The multi-layered feature of the house can even be spotted in the fencing.
2 Looking up from the northern side. The diagonal glass panel slips out of place against the concrete construction.
3 The details of the facade on the road-side.

2

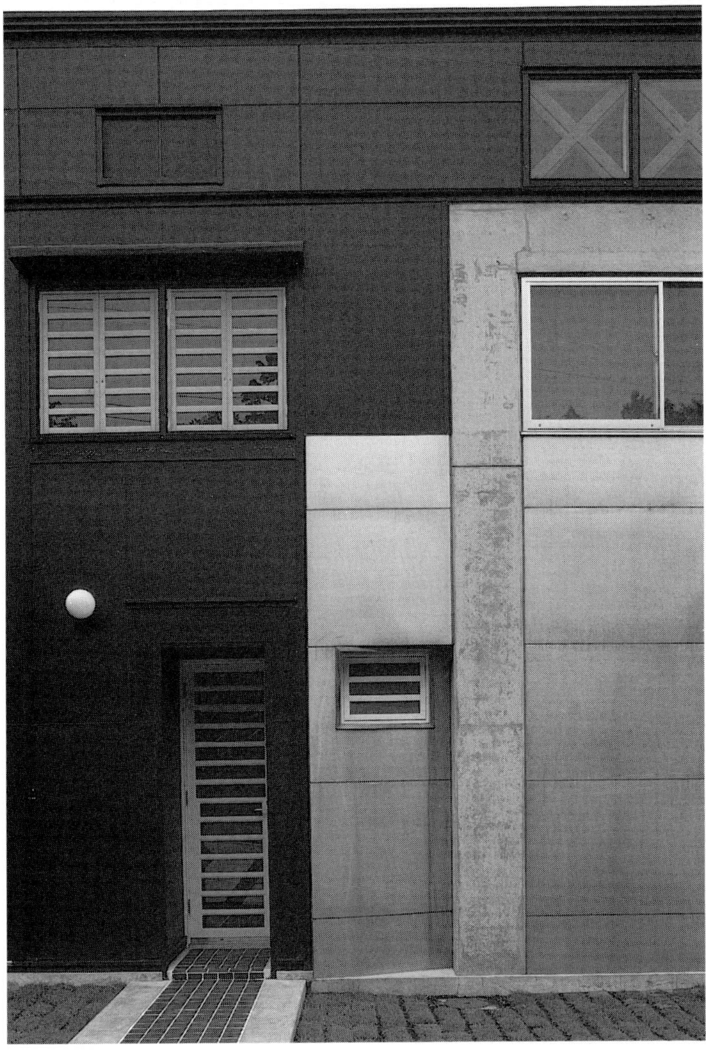

3

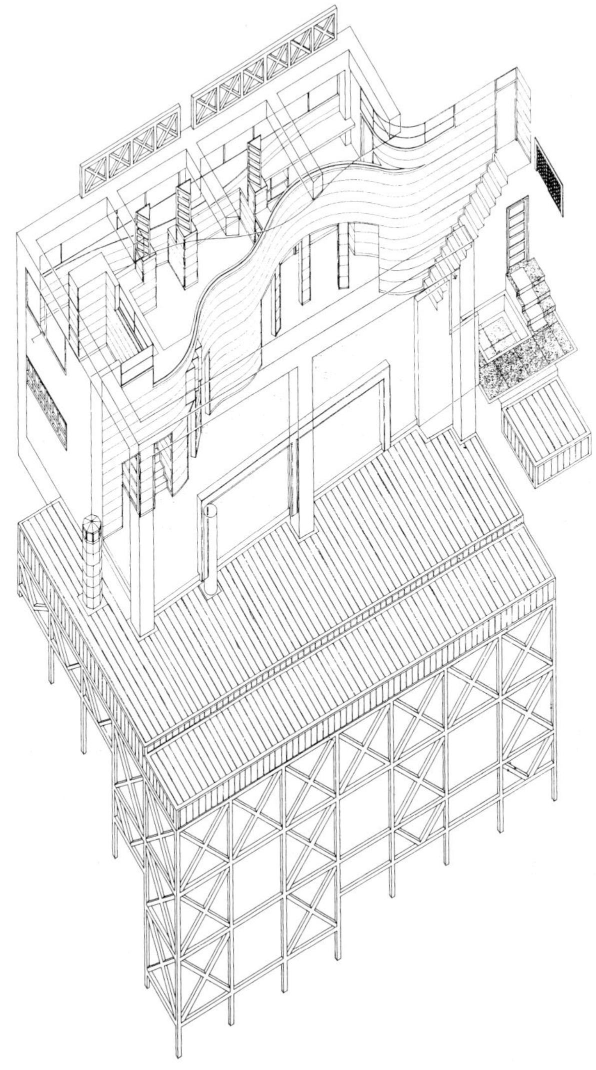

4

5

6

7

4道路側ファサードの詳細。コンクリートの柱と、カーブを描くステンレス板。微妙に角度を変えて立つ4つの扉。
5テラスから居間方向を見る。ガラス面は、左手から右手へ微妙に斜めに後退し、右端に雨戸用の戸袋を確保する。
6居間に流れ出している曲壁。
7テラス面のガラスの皮膜と曲壁の対照。

4The details of the facade on the road-side. The concrete pillars and stainless-steel panels draw a curve. Each of the four doors at slightly different angles.
5A view of the living rooms from the terrace. The glass panel receds at a slight angle diagonally from the right in order to create the space needed for the sliding door at the back.
6The curved wall flows out towards the living room.
7The contrast of the glass divider on the terrace and the curved wall.

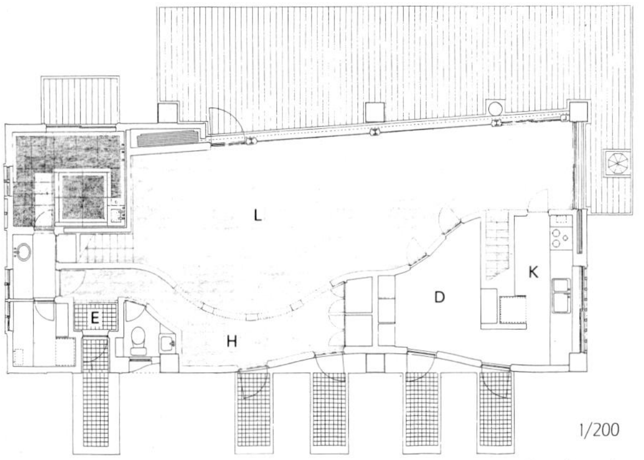

First floor plan

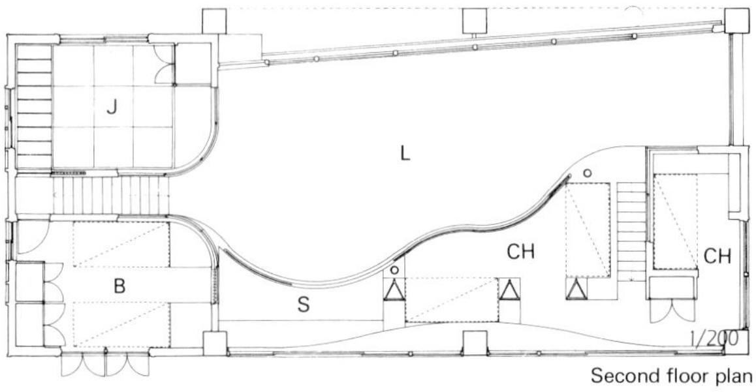

Second floor plan

8

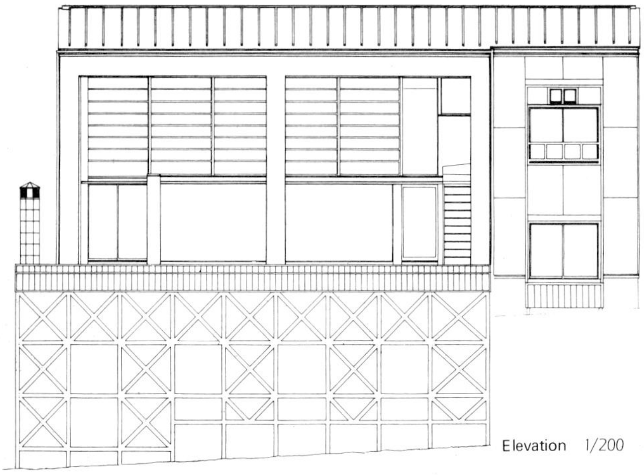

Elevation 1/200

8吹き抜けに開かれている子供部屋。左手前はタワー状の照明器具。
9居間の片面全体に流れている曲壁がつくり出している閉じた部屋は、食堂と厨房だけ。

8The children's room opens to reveal a ceiling that stretches right up to roof level. The tower-shaped lighting can be seen to the left of the foreground.
9The only confined room which is created by the curved wall flowing from the living room, dining room and kitchen.

9

10

11

12

13

10天井を見上げる。コンクリートの骨組が部分的に露出している。
11左に書斎。右手に2階に上がる階段。階段の右に和室、左に寝室。
122階主寝室。閉じた部屋だが、ドアの外の書斎コーナーは吹き抜けに開く。
13主寝室と階段室を挟んで位置する和室。床も窓も和風だが、曲壁とそれに沿った開口部が取り合わされている。

10A view of the ceiling. The concrete construction has been left partly exposed.
11To the left, a study. To the right, a stairway leading to the second floor. To the right of the stairway, a Japanese style room, and to its left, a bedroom.
12The main bedroom on the second floor. Although it is a confined room, the study at the back opens into a roof-high ceiling.
13The Japanese room located between the main bedroom and stairwell. The curving wall and the opening along it have been combined with the Japanese-style flooring and door.

虔十公園林フォリストハウス [Kenju Park Forest House]

■宮城県の北部、市街地から田園の広がる平野を抜けた先、雑木林の丘のなかにすっぽりと埋められたように建っている。それぞれ都心のマンションに住んでいた若夫婦一家とその親の家族が、今までの住まいを処分して森のなかに移住した。
1階に親、2階に若夫婦一家の個室があるが、食堂、厨房、ユーティリティは共同で使う。間取りで見ればコンパクトにまとめられたこれらの部屋の外側、テラスのように開かれ、不整形な形状のフロアが居間となっているが、ほとんど半戸外といっていい演出の仕方である。大人たちはだんろを囲んで集まり、あるいは中庭に面したコーナーを選んで落ちつき、子供たちはその間を縫って、自転車を走らすといった使われ方が想像される。2階のバルコニーと呼ばれる幼稚園の遊戯場のような、曖昧な広がりをもつ場所も、子供たちのもうひとつの遊び場所になるだろうし、そこにも大人が関わってくるように、書庫のコーナーがある。
コンクリートの列柱が、家のなかにもうひとつの林を再現している。この「木」のモチーフのエレメント以外は、不規則な形――下草や枝や梢や雲の形象を真似る。「多層構造」の手法は徹底し、曲折したガラス壁に映る虚像までもが、もうひとつのスクリーンとして利用される。

1

2

3

■This house is situated on a hill in the midst of a copse in the northern part of Miyagi prefecture. The plain that stretches from the city seems to bury the area in remote countryside. The young couple who live here with their parents have sold their condominium in the city and moved out to the woods.
The parents have their own room on the first floor, and the young couple on the second floor, however, the dining room, kitchen and other utility rooms are common ground. The corridors that link these compactly arranged rooms open out like a terrace. The room with the uneven floor is the living room. The overall design is that of a semi-outdoor feel. A clear picture of children riding bicycles around the house while the grown-ups gather around the open fire that faces the courtyard can easily be conjoured up. The balcony on the second floor is more like a playground and provides plenty of room for youngsters to play. The bookcase has been placed there so the adults can also enjoy this space. The lines produced by the concrete pillars bring the sense of another forest into the house. Apart from the wooden motifs, all the shapes are irregular to imitate the swaying of the grass, branches, twigs and clouds. The feeling of multi-layered calmness has been achieved here. Even the vertical images reflected in the mirror can be used as another screen.

1アプローチの坂から見る。2階建ての個室棟と、低層部(居間)との組み合わせ。正面に玄関。
2庭側から見ると居間の壁の表情は一変する。
3連続するガラス壁の一部が切り込まれて、玄関になる。
4打放しコンクリート、フレキシブル板、鉄板、ガラスの対比。

1A view from the approach of the hill. The combination of the two-story structure containing the private rooms and the layers of the lower area. The entrance is at the front.
2The wall of the living room changes completely if viewed from the garden.
3A part of the glass wall produces the entrance.
4The contrast of the ferroconcrete, the flexible sheeting, the iron sheeting and the glass.

4

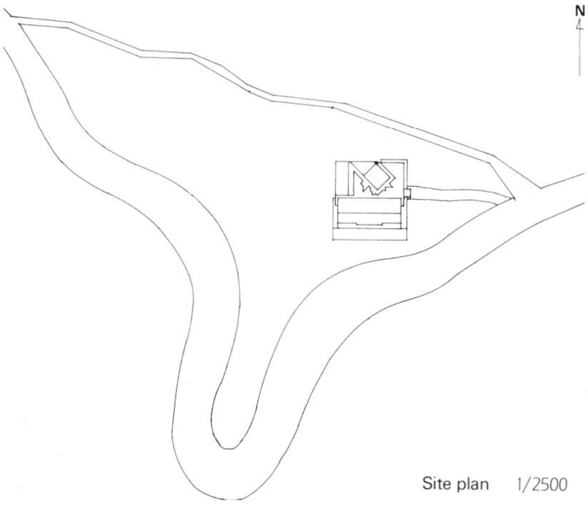

Site plan　1/2500

5

6

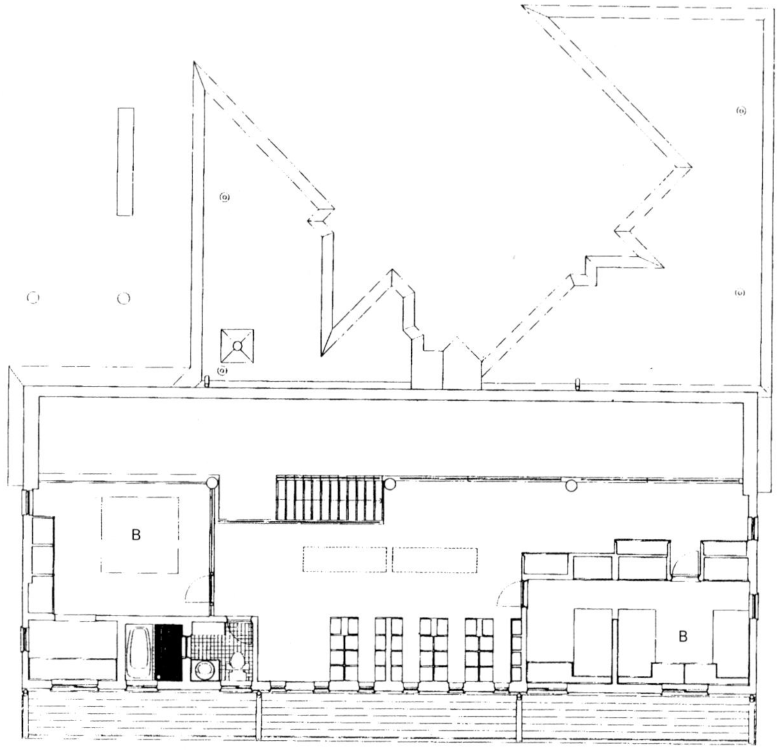

Second floor plan

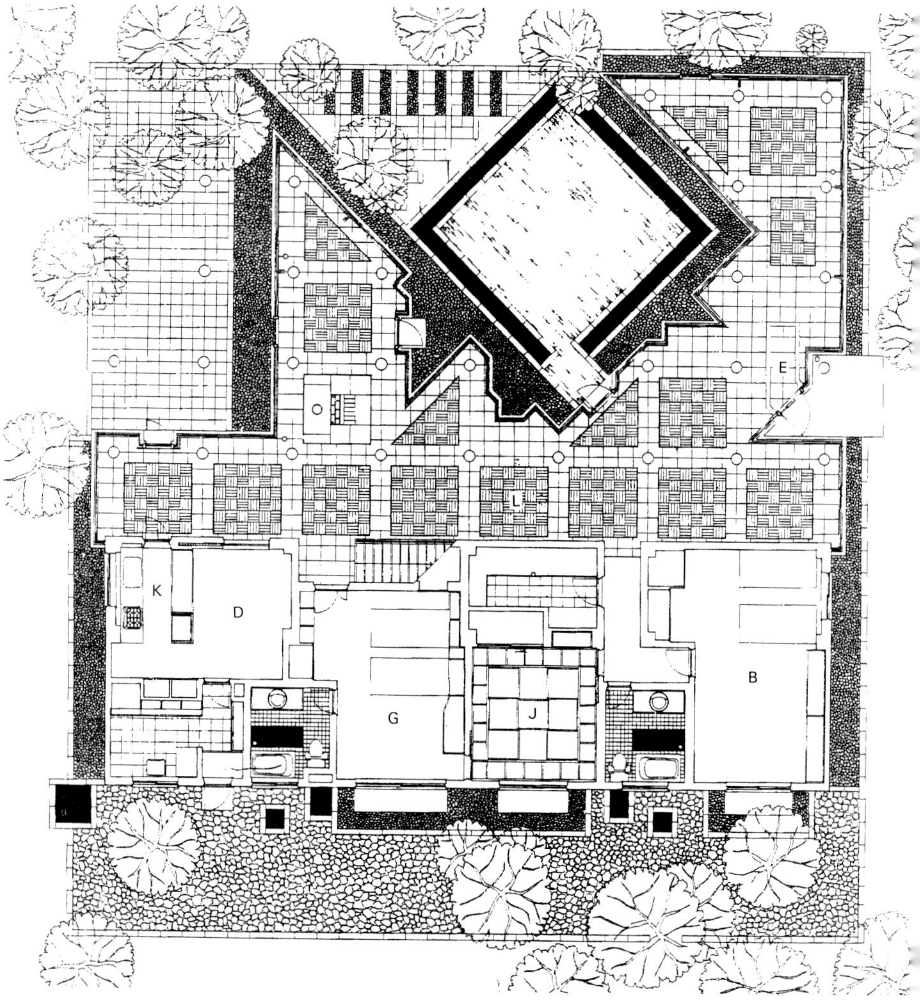

First floor plan　1/200

5居間コーナー。といってもガラスで囲い込まれたテラスというべきだろう。
6コンクリートの柱と大梁の上に、木の梁がかけられている。幹と枝と梢の姿。
7コンクリートの壁で固く囲われた個室とその外側に広がる居間の不整形な表情。

5A corner of the living room ; or should it be called a terrace confined with glass?
6The wooden beams have been built over the concrete pillar and main beam. It is a caricature of bark, branches and twigs.
7The private room surrounded by concrete and the odd-shaped living room that spreads out beside it.

7

8

9

8東西に長く走るコンクリート梁と対照的な雲形の天井は、ラワン合板1枚貼りでつくられている。
9反射像のようにも見える意匠の、手摺の飾り金物。
1011居間の一隅に据えられた暖炉。床まわりと上部は御影石、内側は耐火煉瓦。

8The cloud-shaped ceiling contrasts with the concrete beam that runs from east to west and is made of lauan.
9The decoration of the railings seems to be the design of a reflex image.
1011The open fire in the corner of the living room. Granite has been used for the base and overhang, and fire-resistant bricks for the inner part.

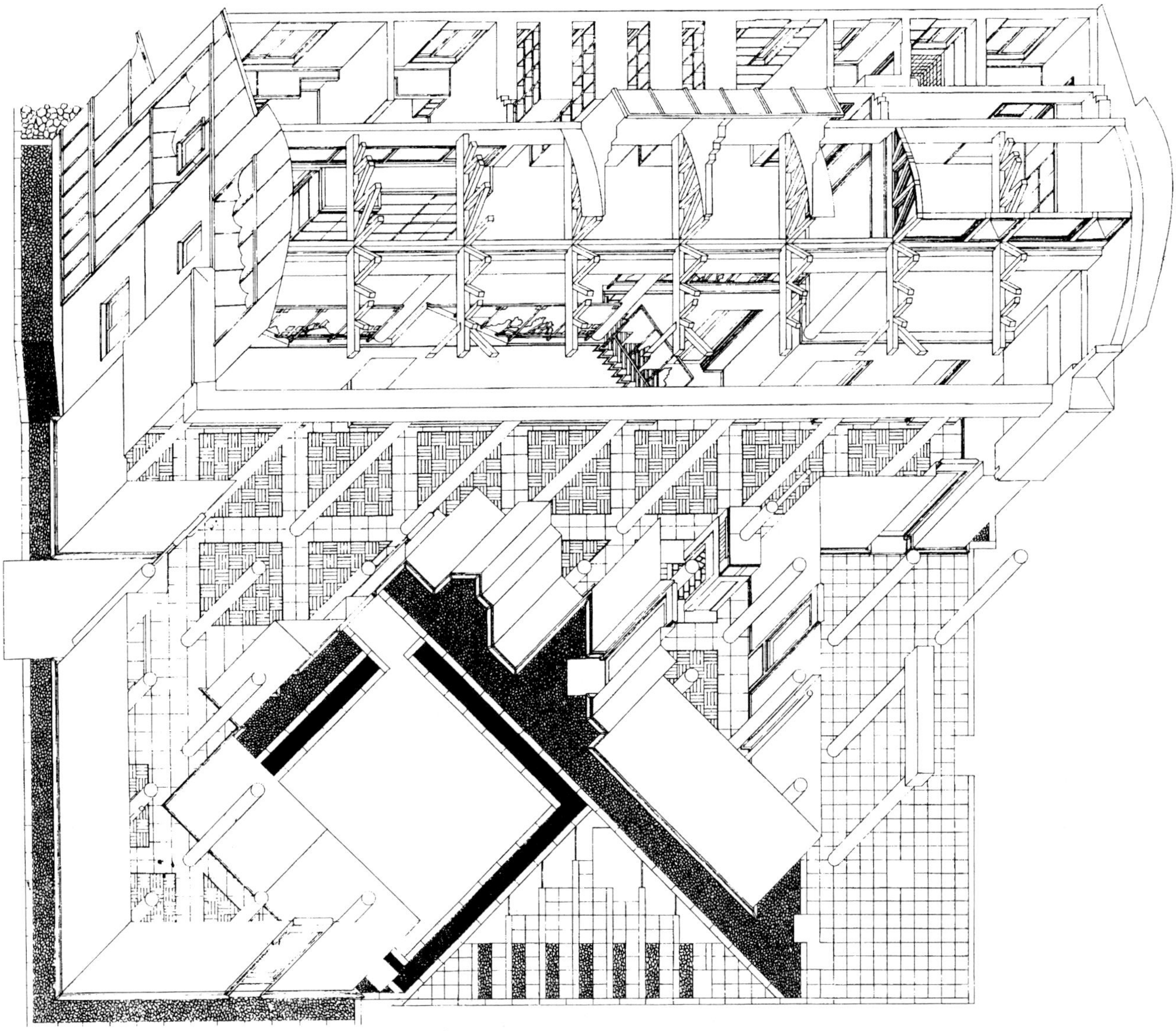

10

11

12

13

14

12中庭を見る。ガラスにさまざまな図形がスリ加工されている。現在は雪の積もっている中庭は、玉石敷き、玉石コバ立てに芝生で整えられている。
13居間の床はテラゾー・タイルとブナ・フローリングブロック。無彩色どうしの構成。
14実景の林と、映像の林の隔合。
15階段の手摺と木の梁、それぞれの表情。

12A view of the courtyard. Various figures can be seen reflected in the glass. At the time of the photograph the courtyard was covered with snow, so the only arrangements are boulders and the lawn.
13Terrazzo tiles and beech flooring have been used for the living room. The construction of neutral colored materials.
14A fusion of two forests ; one real, the other artificial.
15The banisters and wooden beams.

15

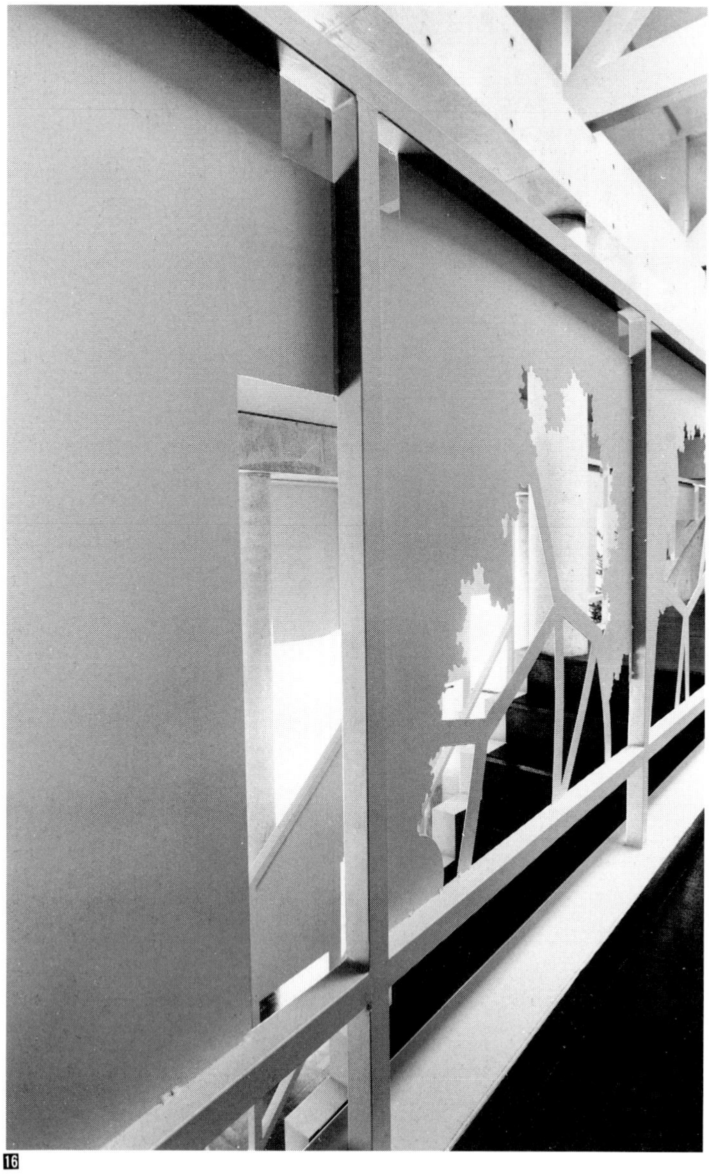
16

17

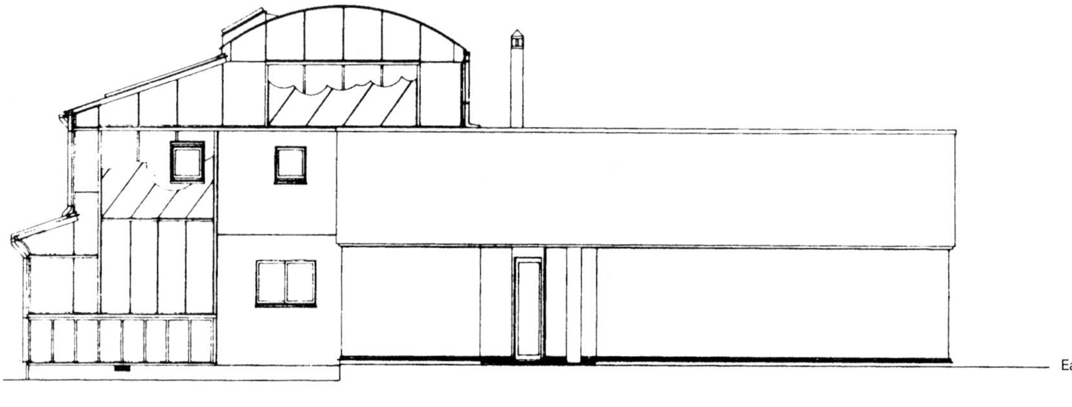
East elevation

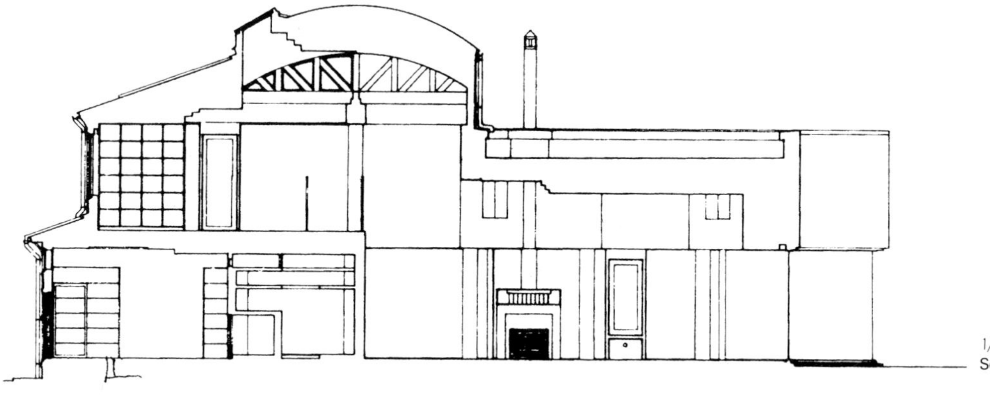
1/150
Section

18

19

16手摺の飾り金物詳細。落下防止の板材が走っている。
17 2階バルコニーに面した寝室。ドアの図形は吹き付け加工でスリ加工に代用している。
18 19思いきり開放的な2階寝室と、しっかり閉じられた1階和室。

16The details of the metal trimmings to the banisters. A metal plate runs along them to avoid them falling.
17The bedroom facing the balcony on the second floor. The finish on the door has be achieved by the use of a spray as opposed to normal methods.
18 19The wide-open bedroom on the second floor and the confined Japanese room on the first floor.

長谷川逸子[HASEGAWA Itsuko]

NCハウス[NC House]

1

1住戸内部。居間。スカイライトは、アルミ・パンチング・メタルとガラスでピラミッド形につくられている。
2各住戸の南面にとられた開口部と、それに続くテラスの表情。

1Inside the apartment. The living room. The skylight is made of pounded aluminium and glass in the shape of a pyramid.
2The southern opening of each apartment and the terrace that continues from them.

2

3

3北側より入口部分を見る。シリンダーの柱は、最北端の住戸の半円形のアルコーブがそのまま外部に現れたもの。角柱は階段室。ガラス・ブロックがはめ込まれている。

4各住戸の東面。三角形を上下に組み合わせた窓とそれに続くテラスの円形の開口部。

3A view of the entrance from the northern side. The cylindrical pillar is an extension of the semi-circle alcove that the house of the northern end has.

4A view from the east. The windows are based on a design of two triangles, which contrast well with the circular openings on the terrace.

■長谷川逸子のほぼ完全なコンクリート住宅は13年前の「緑ヶ丘の住宅」だけである。しかもこの一軒も、木造で計画されたものが、施主の希望でそのまま打放しコンクリートの住宅に置き換えられたという。彼女の住宅は木造、鉄骨造主体のものがほとんどで、それはつねに多層でしなやかな、箱的であるよりは衣裳に近い建築への志向から求められている素材であり構造であるからだろう。

それだけに、基礎あるいはそれに準ずる低層部に用いられるコンクリートの壁体は、全体としては異形に近い姿をしばしば見せる長谷川の建築が、きわめて健実な性格に基づいていることを示す。例えば最近の、「練馬の住宅」や「東玉川の住宅」においては、土—波あるいは雲—星といった垂直方向の位相のなかで、打放しコンクリートとパンチング・メタルといったコントラストの明快な素材が、一体的な家のイメージから宇宙的なダイナミズムへと、建築を拡張している。その展開は、平面計画的にも囲繞的な、つまり宇宙的なまとまりへと、住宅を変貌させていく。「練馬の住宅」においては向かい合う2家族の家の屋根が太陽に向かって伸びる芽のような曲線で、2棟を結ぶ雲の階段と屋根に連続する。「東玉川の住宅」では、満ち満ちた自然——光、植物、水を支えきれずに、縁の外まで溢れさせているコンクリートの器がある。

家が家族を閉ざすものとは反対の表情。

■特異な平面の形。建築家自身による説明では、厳しい外的条件——法的容積制限いっぱいの面積、東のバルコニー側に2m空けられた避難道路、法的採光距離を確保するために45度にふられた壁面、などから自ずと建築の形態が決定されていったような印象を受ける。

しかしそれだけでは、これほどまでユニークで魅力的な集合住宅は出現しないだろう。鋸状の半透明なテラス、シリンダーやたくさんのピラミッド形の屋根、顔を見合わせている角柱と円柱。こうしたものが集まった光景は童話的ともいえるが、それは何気なく重ねられた積木のようにプリミティブで愛らしい。

コンクリートの躯体は、直線の通路とそれに房状にとりつく変形プランの住戸というきわめて単純な構成だが、各住戸の端にとりつけられた、平面で見ると爪のような三角形のテラスを囲うアルミのパンチング・メタル、その透かし模様のような円と四角の開口部、3階の住戸に3つずつ乗せられたメタルとガラスのピラミッド形の屋根など、付加的なエレメントがむしろ主調となっており、コンクリートの躯体はそれを補完するサブシステムとして見えるように扱われている。

入口部分で思いがけないスケール感をもつ角形とシリンダーのコンクリート柱は、閉じた壁面だがユーモラスな表情を帯び、東・南面の開いたテラスが逆にプライバシーを守る。この転換は、集合住宅のなかに見えない中心を想像させる。

4

■The only concrete constructed house that Ms. Hasegawa has designed so far is the Midorigaoka House which was built thirteen years ago. Originally even this house was planned in wood, but it was changed to ferroconcrete at the request of the owner. Most of her houses are based on wood and steel-frame constructions. This is owing to the fact that the materials of construction are usually the decision of the architects who wear them like a costume and enable them to produce multi-functional and flexible boxes. The concrete walls which are used foundation or core of the construction, indicate that Ms. Hasegawa's architectures, are often represented as oddshape, found on her reliable characteristic.
For example, in the case of the Nerima House and the Higashi Tamagawa House, the architecture has been extended passed the field of usual design and has reached the dynamic field of space which has been achieved by the use of contrasting materials such as ferroconcrete and pounded metal to produce a link between the waves of earth and sky. Such developments figure in the basic planning of the construction as the arrangement of the space is being decided. In the case of the Nerima House, the roofs of the two opposed houses continue to the staircase and link the two buildings with curves that appear like buds growing in the direction of the sun. The Higashi Tamagawa House has a concrete containment that overflows with a wealth of nature as it cannot hold all of the plants, water and light. The expression is to hold the interest of the family within the house.

■An unusual shape to the ground plan. According to an explanation received from the architect herself, it was built under the strictest of conditions to meet exact legal requirements. The area used matches the minimum space allowed by law, the two-meter access road has been created on the east balcony side and the wall placed at a 45° angle in order to maintain the legal distance from the lighting. Despite this, it seems as if the house has been designed quite naturally.
However, it takes something extra to be able to create such an attractive and unique house; the saw-toothed semi-transparent terrace, cylinders and the use of pyramidal roof, the square and circular pillars facing each other. Such imagination is usually only available in children's story books. Although it is nothing like the post-modern architectural style and seems slightly primitive, it is a cute as a pile of casually strewn blocks.
The construction of the concrete body is very simple. It is a structure that has been pinned onto the alley like a fringe. However, the emphasis here is more on the pounded aluminium rails, the square and circular shapes, the glass pyramid roof as opposed to its concrete shell. The concrete body has been traded off as a sub-system compromise.
The square and cylindrical concrete pillars that give off the impression of unexpected scale are closed-wall surfaces, but carry with them a touch of humour. The terrace, which opens to the east and south, manages to maintain privacy. Such diversity will give one an idea of what the interior is like.

5

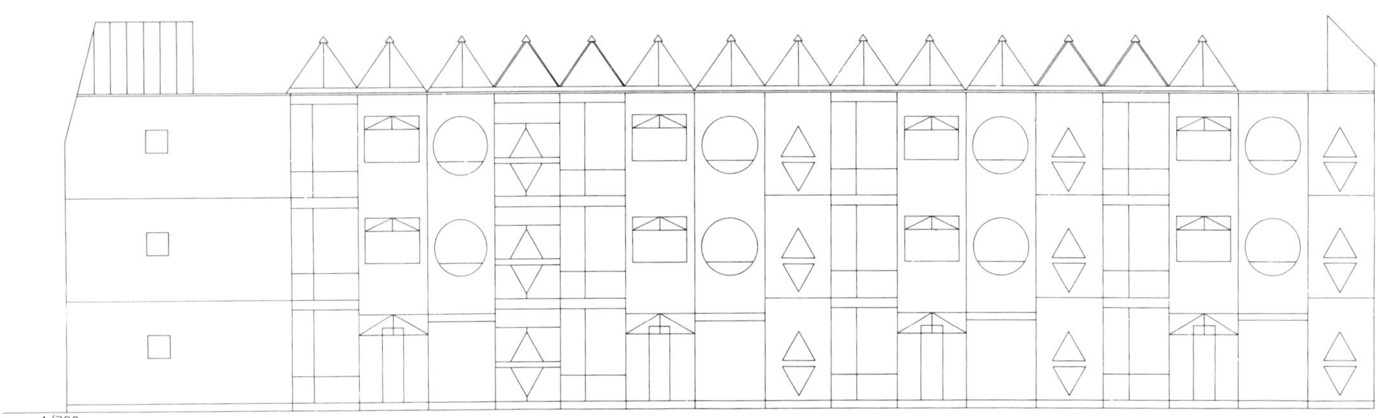

1/200

An expansion of elevation

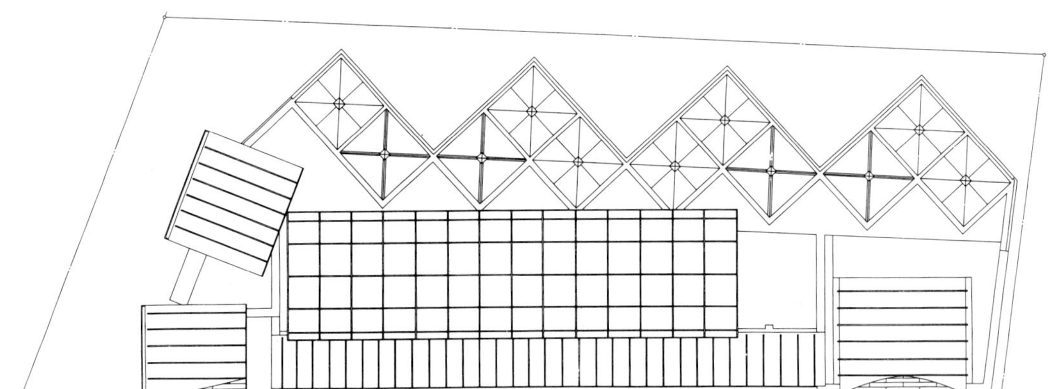

Rooftop plan

6

7

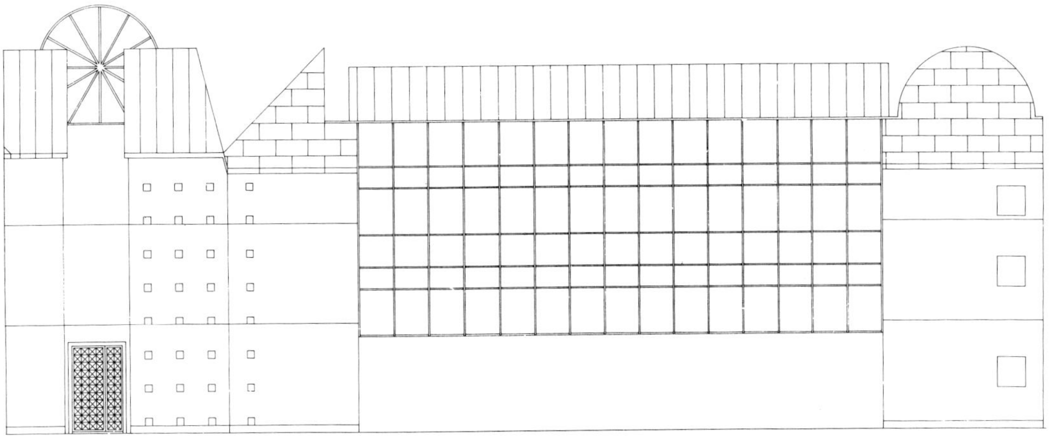

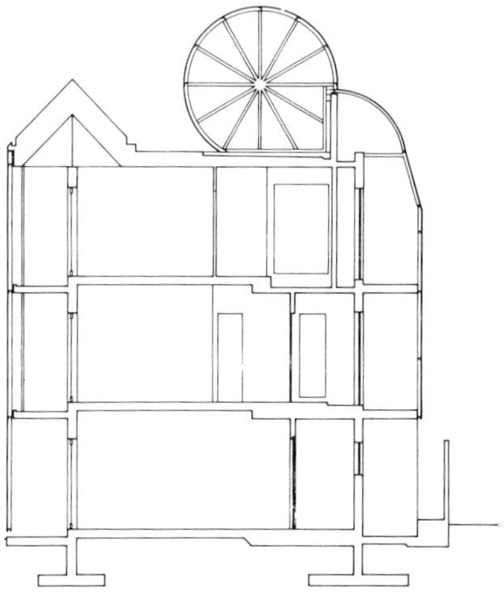

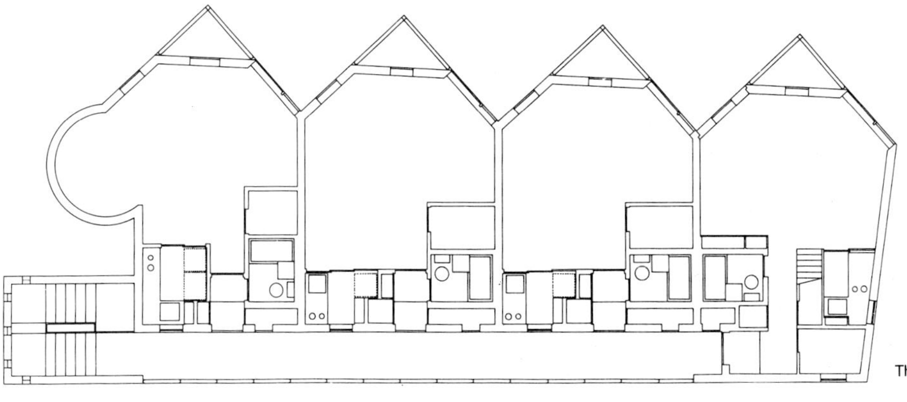

1/200
Third floor plan

5 1階住戸の足元。2、3階のテラスのアルミ・パンチング・メタルの腰壁が、1階では植込みに代わっている。
6 三角形に突き出たテラスの状態がパースペクティブに見える。
7 西側、3階の通路。

5 The foot of the first floor. The pounded aluminium wainscotting of the terrace, second and third floors have been replaced with plants on the first floor.
6 A perspective view of the triangle-shaped terrace.
7 The west-side passage on the third floor.

8

9

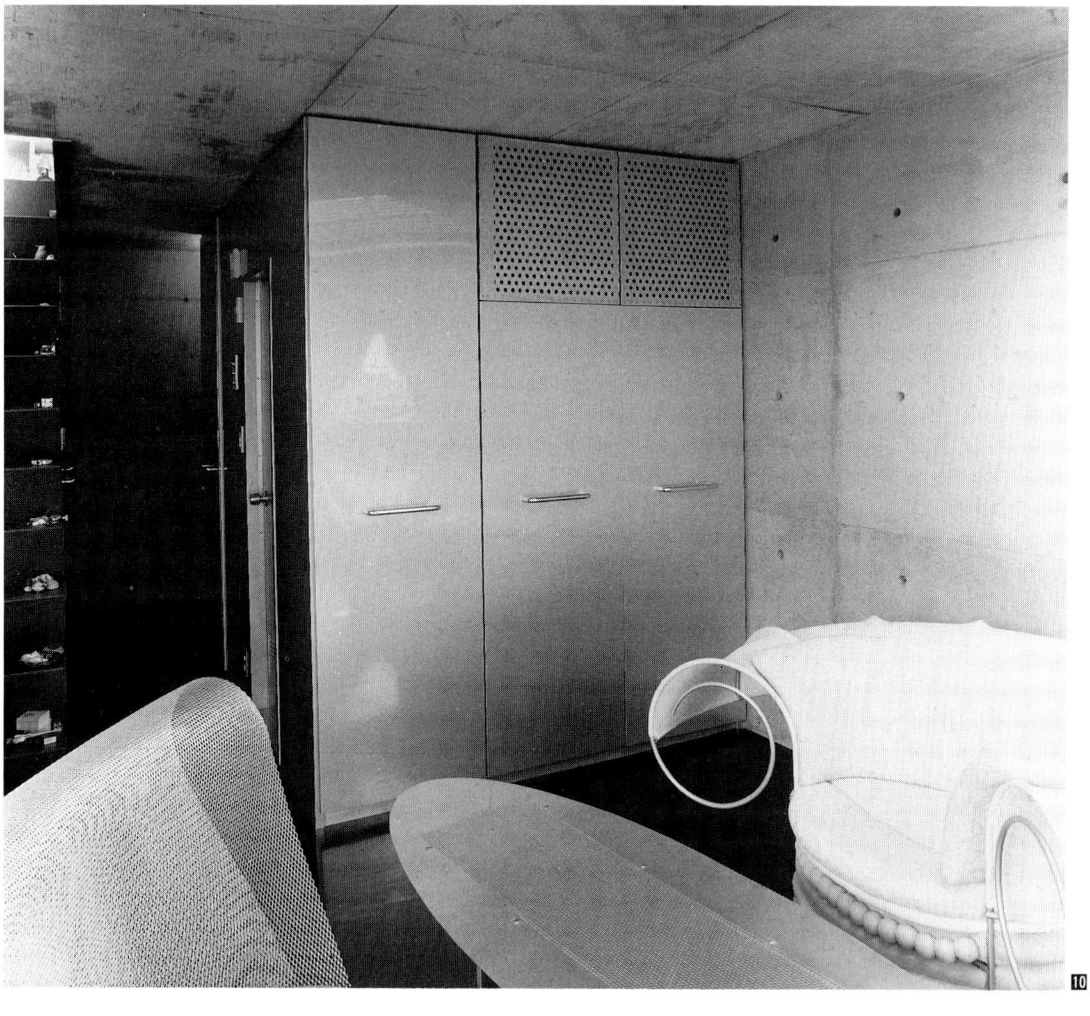

10

8階段室。
9 10 11住戸(3階南端)内部。

8The staircase.
9 10 11The interior of the apartment (southern end on the third floor)

11

12

13

14

12天井を見上げる。3つのピラミッドのうちの1つがスカイライトになっている。
13シリンダー形の上階内部。
143階各住戸に3基ずつのピラミッドの屋根。将来はそのひとつが、手前のスカイライトと同じものに変えられる予定。

12A view up to the ceiling. One of the three pyramids is a skylight.
13The interior of the cylindrical upper floor.
14The three pyramids on the third floor. In the future one of them will be changed into a skylight like the one in front.

早川邦彦［HAYAKAWA Kunihiko］

STRATUM

■早川邦彦は、積極的にコンクリートを用い、しかもこの一種類の建築素材に、できる限り多様多弁の言葉をしゃべらせようとしている。
例えば、成城学園前に建つ3軒並びの住宅(はじめから一度に計画されたのではなく、80年から8年間の間に同じ親族によって順次依頼された、日本ではまれな例)は、いわば何層にも増殖分化した、打放しコンクリートの外皮をもつ。この外壁は、格子状、段状、円、半円等のさまざまな開口部で開かれて、多層の状態を演劇舞台のように見せている。あるいは「市川大野の家」のスタジオ部分や中澤ホールの入口まわり、あるいは最近完成した「アングル」の各所に見られる、グロットの岩か崩れた廃墟のような、断片的で不規則な凹凸の壁。
早川の建築には、こうしたヴォキャブラリーのとめどない「氾濫」があり、その勢いはさらにスチールやガラス・ブロックなどの素材まで引き寄せている。この氾濫をいっそう掻き立てているように見えながら、じつは建築の新しい秩序や意味をつくり出しているのが、コンクリートやスチールに思いきり数多くの種類の色を塗ること、という手法である。それは、建物と生活する部分とを分けて可視化し、建築の外部と内部という概念に、新しい切り口を示したのである。

■「ストラタム(都市の地層という意味の命名)」は、早川の建築群のなかでももっとももささやかなもののひとつだが、同時にここ6、7年、早川がアイデンティティを確立してきた表現圏を、もっともよく見晴らせる地点に建っている。
例えば「アトリウム」においては、思いがけない形態と色彩のコラージュによる仮面の中庭と、打放しコンクリートや植栽や噴水による自然の庭との対比が、個々のユニットの集合体を示すという集合住宅の常套的な構成手法を裏返しにしているが、この新しい雑居ビルでは、対比的に構成されていた外と内がよりいっそう近寄り、ついには癒着してしまう。ポーラスな打放しコンクリートの壁と彩色された壁が重なりあって1枚になり、あるいは単純な壁とシリンダーの屋根や家形の形態が一体化する。1階が店舗、2階がオフィス・スペース、3、4階はメゾネット・アパート。周囲も商店や住宅が混在する環境である。80年代初期の「南青山の家」から試みられていた、街を家に映し込むという着想は、この建築においてより端的、より適確な表現を得るに至っている。それは見え掛りとしての表現から、外皮と内容、表と裏といった建築と生活に同時に関わるコンセプトを引き出す次元に及んで、はじめて説得力をもちえたともいえる。
こうした構成手法は、その後に設計された「ステップス」においてさらに劇的に展開されていくが、「ストラタム」はその一転換点である。

■Mr. Kunihiko Hayakawa invariably uses concrete in his houses. He practically manages to make this hard medium speak. For example, it is worthwhile pointing out the row of three houses built by Mr. Hayakawa in Seijogakuen-mae (none of these houses were built at the same time. Commissions from the owners were received on three separate occasions over a period of eight years starting in 1980. This is an unusual example indeed). They are all constructed in layered ferroconcrete. This external wall has been constructed upon the lines of a theater stage by creating various openings which seem to conjure up the feeling of the stalls and circle. The whole structure is slightly fragmented in the same fashion as a grotto or demolished ruin, and this sense is especially strong in the studio part of the house in Ichikawaono, around the entrance of Nakazawa Hall and in various areas of the "Angle", which has recently been completed.
The architecture of Mr. Hayakawa contains an eternal flood of varied vocabularies which give strong influences to the inanimate materials like steel, glass and bricks, etc. In fact it appears that recently these vocabularies are being stirred up again as new methods and meanings of the architecture are coming to life; such as the painting of the steel and concrete in as many different colours as possible. Such methods have made it possible to distinguish buildings from living quarters and has opened another door in the conception of architecture involving both outdoors and indoors.

■"Stratum (or "layers in the city")" is one of Mr. Hayakawa's more humble constructions. However, it was constructed in perfect timing for him to display his individual expression that he has spent the last six or seven years establishing. For example, as to "Atrium", the courtyard has been disguised to take on an unexpected form, and the colourful collage and the fullness of nature in the garden take on an interesting contrast with the ferroconcrete. The cultivated plants and the fountain turn the common conception of houses being individual units inside out. This miscellaneous building has blended the outside and inside so well that they are practically one. The positioning of the ferroconcrete on top of the coloured wall attains unity, and with the cylindrical roof the formation of the complete construction becomes one.
The first floor is taken up with shops, the second with offices, and the third and fourth floors are devoted to masionettes and apartments. The conception of bringing the town into the house, a conception that was first attempted in the early 80's with the construction of "the house in Minami Aoyama", has achieved much direct and accurate expression with regard to architecture. It could be said that persuasive powers were brought to success for the first time. It has been developed to the stage of extracting the initial concept which is simultaneously related to the lifeblood that is contained within architecture and its contents — both front and back.
Similar constructive methods will be developed further in the "Steps" program which was designed at a later date, but "Stratum" was constructed at the very beginning of this point of change.

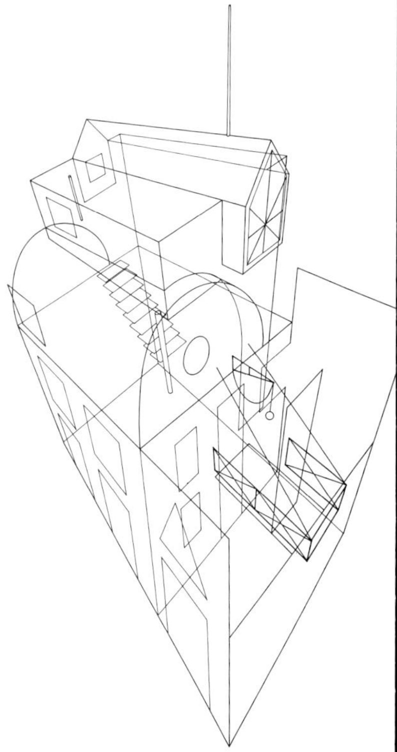

1

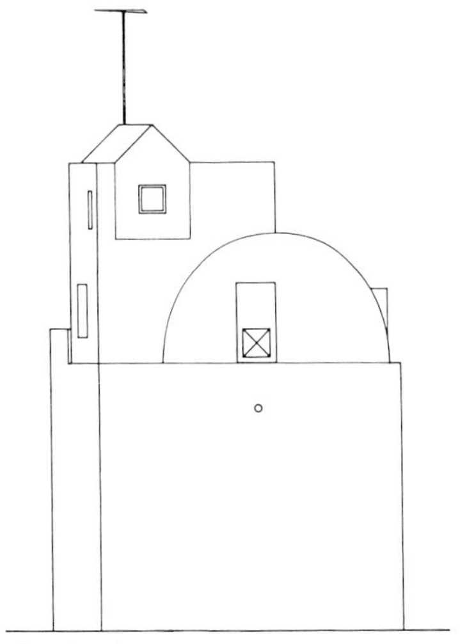

East elevation

West elevation

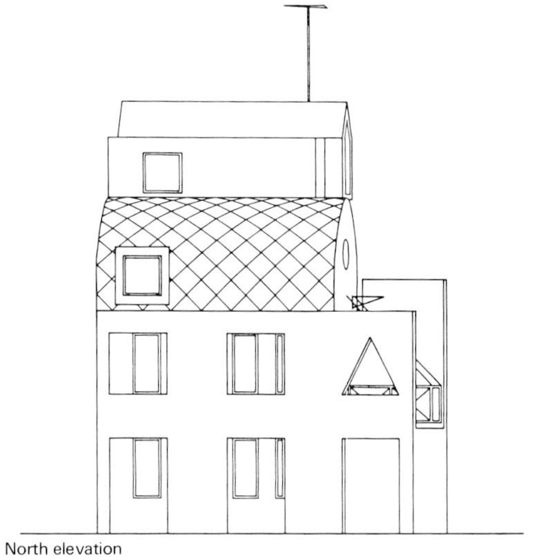

North elevation

3

2

4

1 北側の道路に面して建つ打放しコンクリートの壁がゲイトのように切り抜かれている。前庭を介して左に店舗、奥に2階への外階段。
2 2階から3階へ。建物のなかを貫通した階段は、のぼり切ったところで再び、外気に接する。その左手に3、4階アパートの入口。
3 4 前庭およびその真上の外階段踊り場から表通りを見る。

1 A part of the ferroconcrete wall facing the road on the north side is cut off like a gate. A shop to the left of the garden in front of the building, an outside stairway leading to the second floor at the rear.
2 From the second floor to the third floor. Stairways run through the building and reach the outside air again on the top. An entrance to the apartments on the third and fourth floors to the left.
3 4 The main street viewed from the garden in front of the building and from the landing of the outside stairway above it.

5

6

7

56 3、4階のメゾネット・アパートは、外壁を彩色し、内部は逆に打放しコンクリートと淡いベージュだけで抑えている。外階段や照明器具を支える赤く塗られた鉄のバーが、内部に別のかたちで再び現れる。
7 微妙に変化をつけられた色とテクスチャ。

56 The outside walls of maisonnettes and apartments are painted while the use of color was restrained inside with just light beige and the natural color of ferroconcrete. Red iron bars which support lighting instruments appear again in a different form inside of the room.
7 The subtle change in color and texture.

8

9

10

11

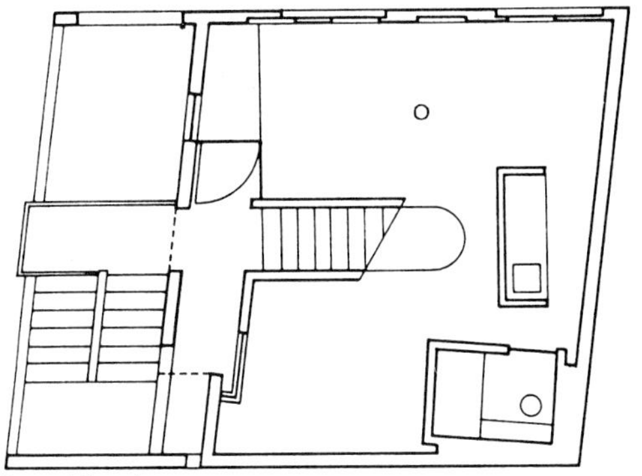

Second floor plan

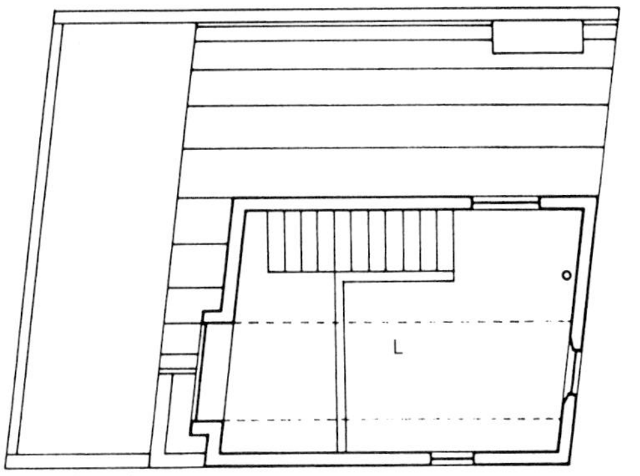

Fourth floor plan

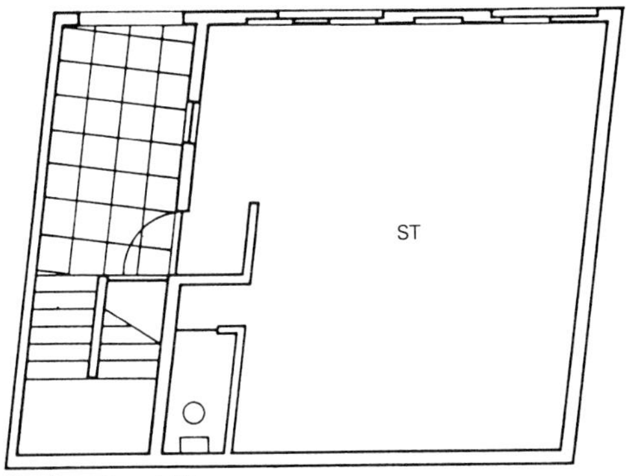

First floor plan 1/100

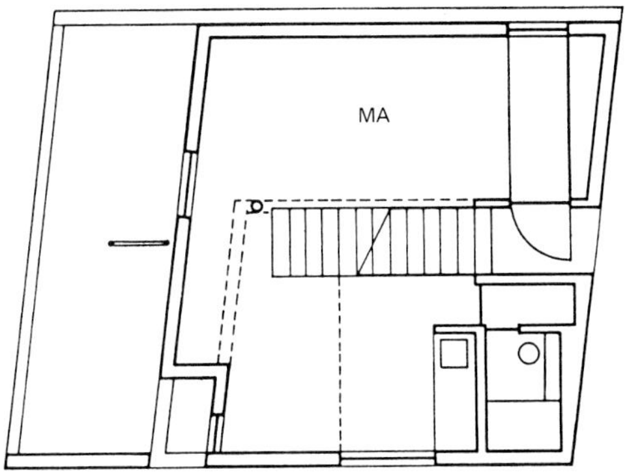

Third floor plan

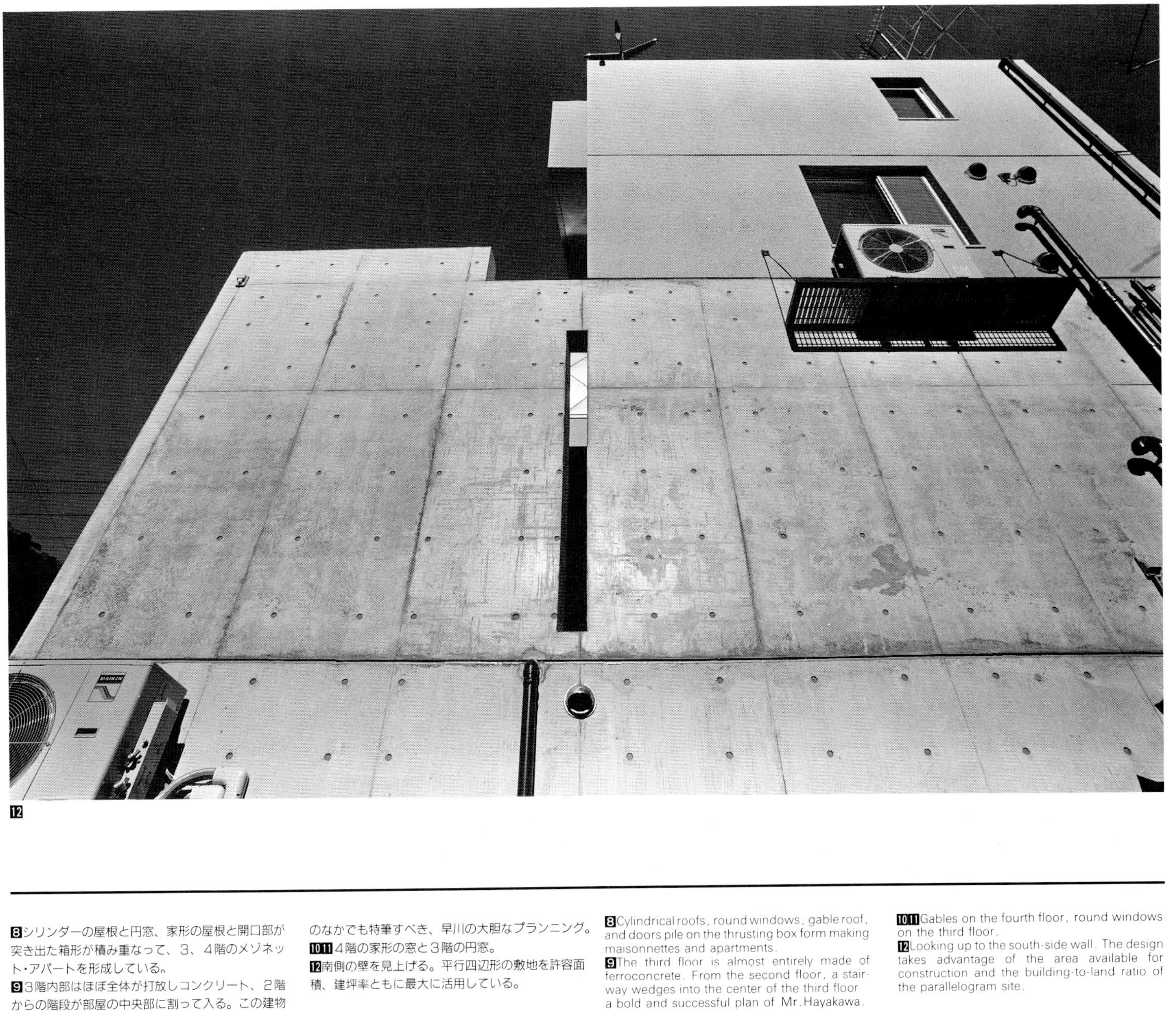
12

8 シリンダーの屋根と円窓、家形の屋根と開口部が突き出た箱形が積み重なって、3、4階のメゾネット・アパートを形成している。
9 3階内部はほぼ全体が打放しコンクリート、2階からの階段が部屋の中央部に割って入る。この建物のなかでも特筆すべき、早川の大胆なプランニング。
10 11 4階の家形の窓と3階の円窓。
12 南側の壁を見上げる。平行四辺形の敷地を許容面積、建坪率ともに最大に活用している。

8 Cylindrical roofs, round windows, gable roof, and doors pile on the thrusting box form making maisonnettes and apartments.
9 The third floor is almost entirely made of ferroconcrete. From the second floor, a stairway wedges into the center of the third floor a bold and successful plan of Mr. Hayakawa.
10 11 Gables on the fourth floor, round windows on the third floor.
12 Looking up to the south-side wall. The design takes advantage of the area available for construction and the building-to-land ratio of the parallelogram site.

13

14

15

16

17

18

13 14 15 コンクリートのさまざまな面。それにパンチング・メタルのドア(ポンプ室)や赤く塗られた鉄骨の手摺、船舶用の照明器具、屋上の白鷺をモチーフにしたオブジェ(山本衛士作)などがアクセントになっている。山本のオブジェは4階オフィスのテーブルにも見られる。

16 開口部から見える外の風景が、もうひとつのテクスチャとなっている。

17 4階リビングの東側の窓。

18 内階段より2階外階段のほうを見下ろす。左右の三角形のガラス越しに3階の室内。

19 20 道路からの全景。打放しコンクリートをそのままに、あるいは平滑な、あるいは手触りのある仕上げ材を用いて、多様な色とテクスチャを構成。内外で約20色を使用。

13 14 15 The various faces of the concrete ; the pounded metal door to the pump-room, the red-painted steel-framed railings, the boat lighting and the objet d'art motif (by Mr. Eiji Yamamoto) of the heron on the roof-garden have all become accents. The objet d'art by Mr. Yamamoto can also be seen on the table in the fourth floor office.

16 A view from the door adds yet another texture to the room.

17 The fourth floor living room window facing east.

18 A view of the outside stairs from the inside stairs. A peek inside the third floor through the triangle glass.

19 20 A complete view from the road. The colorful and textured construction has been completed with the use of ferroconcrete. A total of 20 different colors have been used for the inside and outside.

19

20

■国立市の住宅街の角地に建つ。敷地内に桜の巨木を残しているが、周囲も木々の多い恵まれた環境である。店舗、貸アパート7戸、オーナーの専用住宅とプライベートなミュージアムとしての屋上からなる、地下1階・地上3階の建物。2つの棟が半階ずつスキップして並び、その間にアクセスの階段が割り込んでいる。それぞれの住戸は桜や周辺の緑の眺めを考慮して、戸建住宅が集まったかのように全く異なる間取りと開口部をもつ。地階の住戸は、この近くにある音楽学校の学生も考慮に入れて、楽器の演奏も可能なように遮音効果が工夫されている。

外壁は、道路側に見せている面は、打放しコンクリートのまま、窓の位置も規則的で、あえて素っ気ない表情にしている。ところが各住戸の入口まわりに近づくと壁はさまざまに彩色され、パイプやガラスが目立ってくる。建物という外皮に包まれた、生活という中味が、びっくりするほど鮮やかに可視化されているのだ。このような集合住宅の新しい建築像はヴォキャブラリーの転換をももたらす。そのもっとも象徴的な例が、この建物の愛称ともなっている階段(ステップス)である。正面階段でもあり生活階段でもある。建築の裏の顔でもあり表の顔にもなっている。桜の木から語り始めた絵巻物としての建築。

■These buildings are situated on a rectangular site in the city of Kunitachi. There is a giant cherry tree to the left, and the whole site is surrounded with greenery. This is a three-story building plus basement which consists of shops, seven apartments for let, the owner's residence and the roof level which is being used as a private museum. The two buildings stand side-by-side with only a half-flight of stairs difference in height. The approach stairs have been constructed between them. Each apartment has a completely different layout of rooms as if they were all individual houses, as they were constructed with a consideration for the views of the cherry-blossom tree and other greenery. The apartments in the basement have been sound-proofed. The local music students can practise their instruments in here.

A plain design has deliberately been chosen for the external wall. The wall facing the road is of ferroconcrete and the windows have been fixed in regular appearance. However, when one approaches the vicinity of the entrance to the apartments, the bright color-scheme and the pipes and glass stand out.

The contents of the potential life-styles available within this complex is refreshingly revealed despite its drab shell. Such fresh architectural images for recent condominiums brings an interesting turn-over in the vocabulary as well. The most symbolic example of these buildings are the steps, after which the construction is named. The front steps and the rear steps. At the same time, the face of the architecture has been cleverly hidden, but also revealingly displayed. This constrcution should be looked upon as a scroll which tells a story; starting with the cherry-blossom.

1

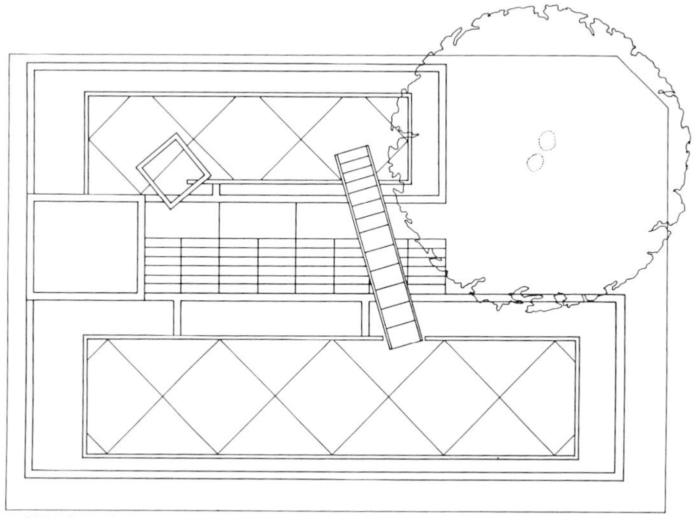

Third floor plan

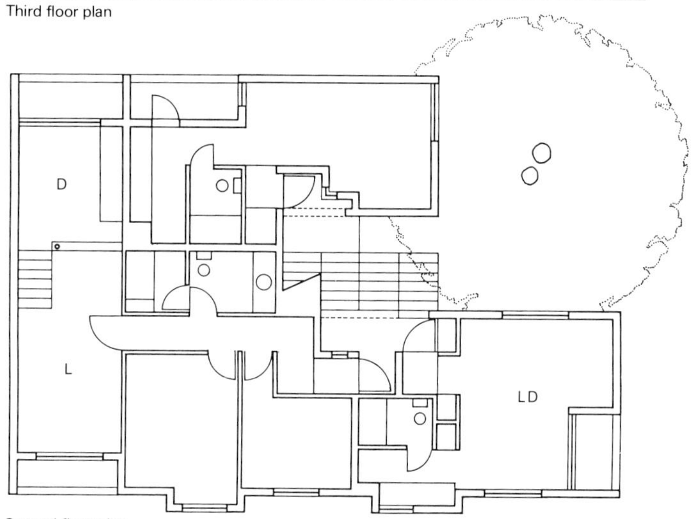

Second floor plan

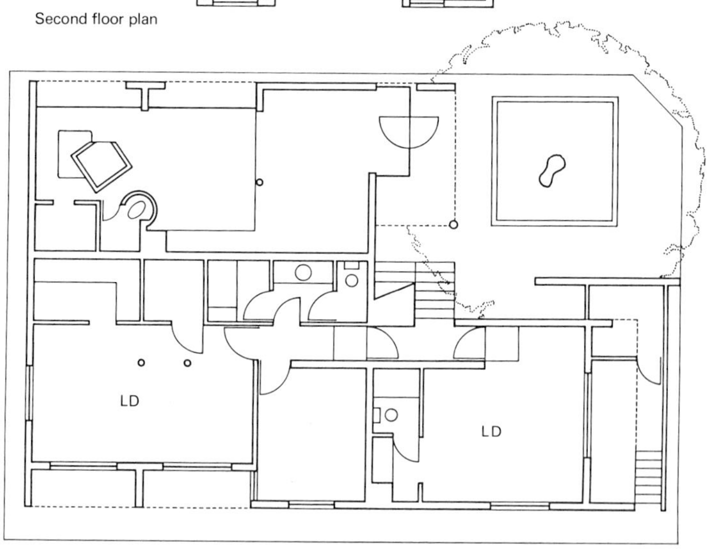

First floor plan 1/200

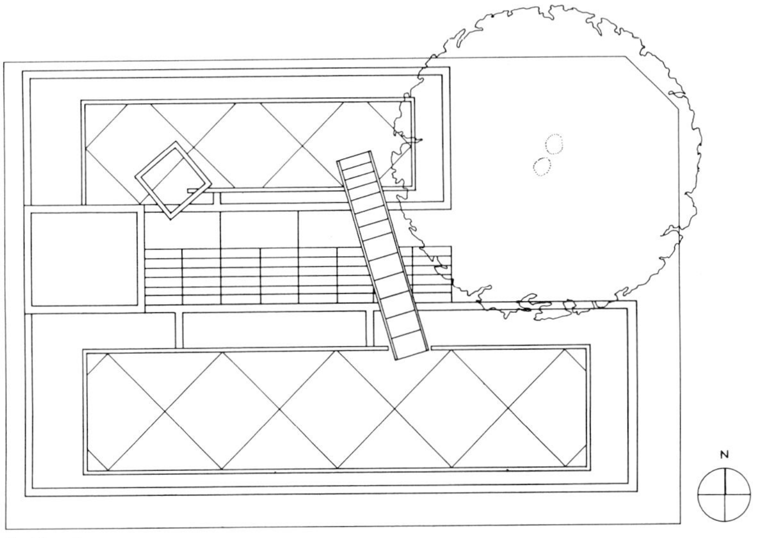

Rooftop plan

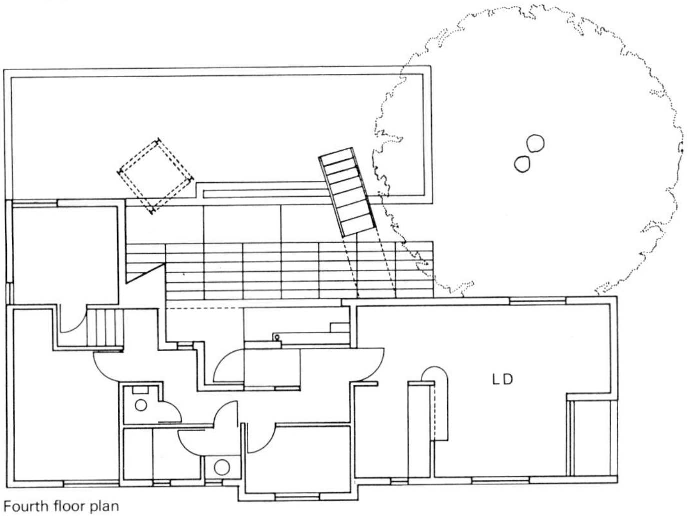

Fourth floor plan

■1 1階ブティック入口からアプローチまわりを見る。
■2 樹齢100年を越えた大島桜が残されている。ここから見る外壁は、ほとんど打放しコンクリートのまま。右手にブティック。左手に半地下階の住戸へのアクセス。

■1 A view of the vicinity of the entrance leading from the boutique on the first floor.
■2 This cheery-blossom is over 100 years old. The external walls in this view have all been constructed of ferroconcrete. The boutique and access to the basement are on the left.

■2

3

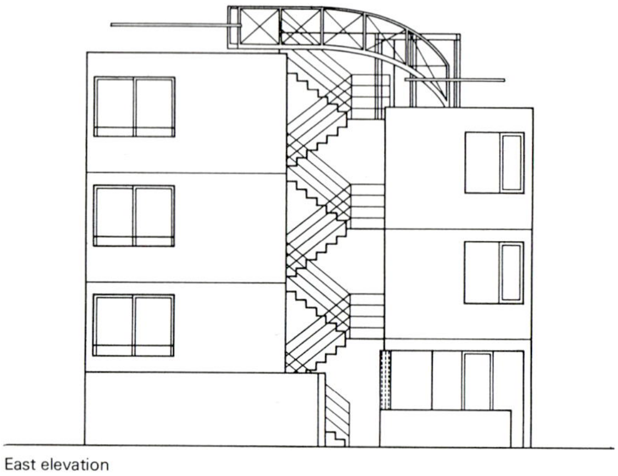

East elevation

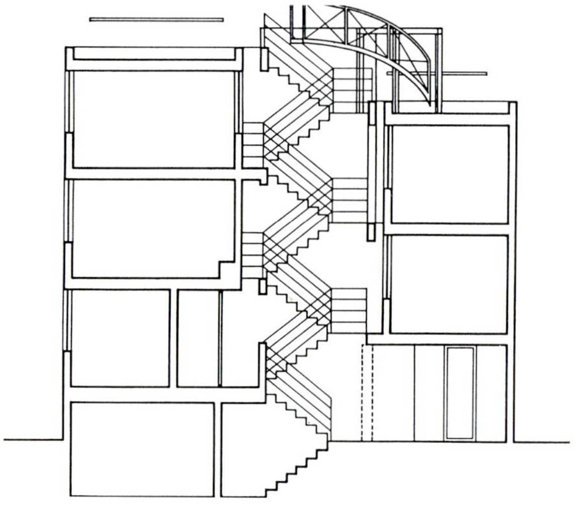

Section

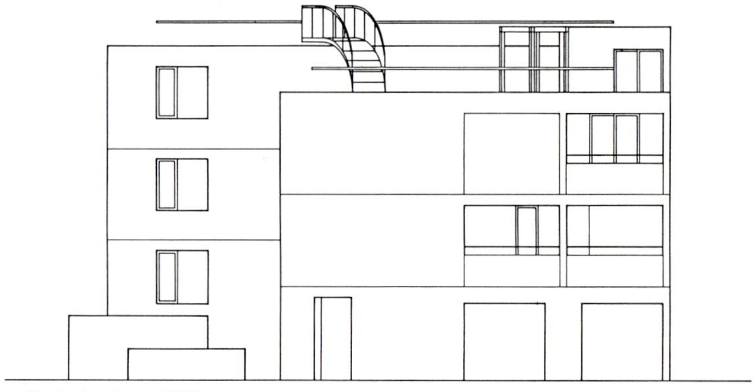

North elevation 1/200

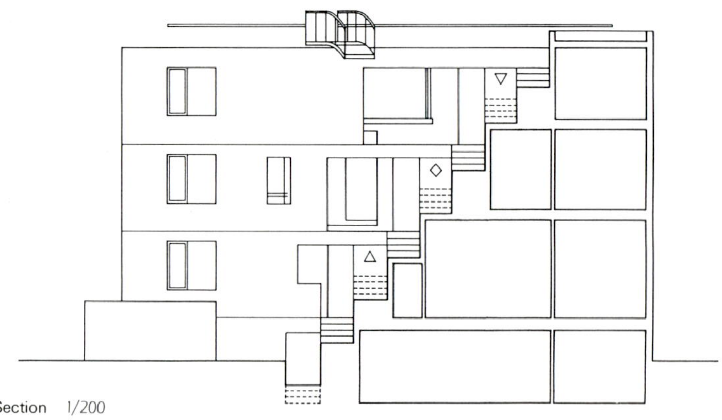

Section 1/200

4

3打放しコンクリートの壁の間から、住戸部分のペイントされた壁や異形の窓が少しずつ顔を出す。
4平行して並ぶ2棟の間に入り込んでいる階段。各住戸へのアプローチ。アンバーローズ色一色にペイントされただけだが、さまざまな色合いに見える。

3The painted wall and odd-shaped widows begin to reveal a little more.
4The stairs built between the two parallel buildings. The approach to each apartment. They have been painted all in amber-rose, and seem to have various color.

5

5打放しコンクリートの外側と、ペイントされた「ステップス」との対照。屋上の、鉄骨を組んだゲイトは黄色。
6屋上から2棟間のスペースを見下ろす。2つの棟は屋上で赤い太鼓橋で結ばれている。桜の梢が真近かに。
7 1階ブティック前から屋上のほうを見上げる。
8屋上の太鼓橋から見下ろす。2棟の間は半階ずつスキップしている。

5The contrast between the ferroconcrete of the outer wall and the painted steps. The steel-framed pergola on the roof has been painted in yellow.
6Looking down the gap between the two buildings from the roof. The buildings have been connected with a red arch-bridge. A close view of the branches of the cherry-blossom.
7Looking up to the roof from the boutique on the first floor.
8Looking down from the arched bridge on the roof. The two buildings are separated in height by a single flight of stairs.

6

7

8

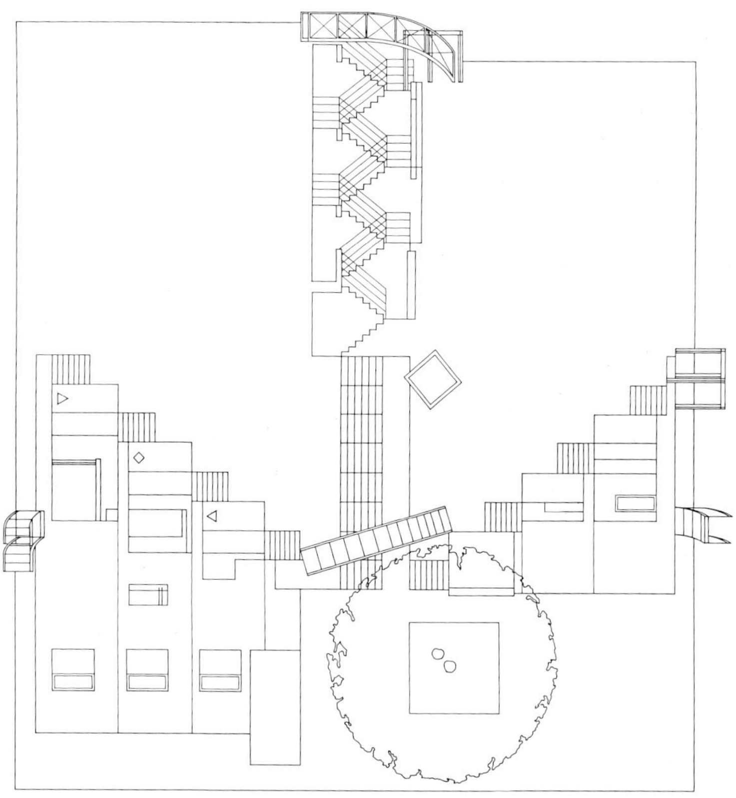

伊東豊雄 [ITO Toyo]

馬込沢の家 [House in Magomezawa]

■伊東豊雄が1970年に、外壁をアルミのパネルでパッケージした、木造の住宅を処女作としてつくり上げたときから、住宅の建築性に対するこの建築家の疑問は、すでに提示されていた。
その後「中野本町の家」「上和田の家」においては、部屋の配列や構造体の序列など、建築を秩序立てて示す要素が極力排除され、伊東が形態素（モルフエーム）と呼ぶ、折れ曲り、あるいは弧状にうねる切れ目なく連続する壁だけにすべてが還元される。天井も壁も床も重力を失って消えていくような流動的な空間は、住まいの場でさえも、単なる体験的なシークエンスを影絵のように集めてしまい、それでもなお住宅として成立させる実験だったといえる。それは一過性の実験にとどまることなく、欧米のモダンリビングの影響をそっくり受けた日本の現代住宅に対する疑問として、より本来的な住宅の形式を発見していく始まりとなった。
この、地下室のように閉じたコンクリートのモノシリックな住宅を経過したのち、ただ床があり、その上部に屋根が浮かんでいるような、開かれたより直截で原形的な家の形へと、一挙に向かっていく。彼の自邸はその典型である。

■およそ8m角のコンクリートの箱があり、その内部と、箱の上と周辺に生活の場がある。その内容はとくに規定されていない。あるとすれば箱の中にビルトインされたキッチンだけである。浴室・便所は部屋というよりは装置として、居室の外側に装着されている。テラスを塀のように囲う倉庫も、さらには大きな傘のような屋根も同じように扱われている。
考えられる限り単純化された住まいであるが、その簡素さが逆にきわめてデリケートなディテールを生み出している。透明な塀と門、玄関と玄関ホール。「スペースA」とヴェランダの一体感。「スペースB」の、室内から室外へ段差をとりながら連続している床。大きく開かれた部屋と強く閉じた部屋。この住宅での日々の生活を具体的に想像していくと、現代の日本の若者たちにぴったりであることがわかる。何もかも切り捨てた裸の建築が、荒々しさへの志向を失い、都会的な洒脱な感性に満ちている。
この住宅は、伊東の師である菊竹清訓の「スカイハウス」を思い出させる。日本の戦後住宅史上のエポックをつくる住宅が、つねに同質の単純さに還ることで出現してきたことにも気がつく。床以外には何もない。キッチンや浴室や倉庫や洗濯槽は、菊竹が提唱したムーブネットに近い。

■Ever since Mr. Ito designed the wood-frame house covered in an aluminium panel for his maiden job, his approach to architecture has never been in doubt.
With regard to "the house in Nakano-honcho" and "the house in Kami-wada" which have been built since then, all of the elements which indicate the methodical aspect of architecture, such as the arrangement of the rooms and the order of construction, have been deliberately illuminated. In addition to this, everything has been reduced to a continuous curved and wending wall which Ito calls Moldform.
This fluidity which seems to make the floor, walls and ceiling lose their weight and disappear into thin air, has the ability to gather together the mere experimental sequences like a silhouette even in a living area. This, in fact, was probably an experiment to prove that it can work on even houses. And the end product did not finish then. It was the start of the era of discovering that modern Japanese houses, with their heavy western-style influence, will respond well to the architectural questions that are asked of them.
Mr. Ito's architectural style has a tendency to side towards open, direct and basic methods; such as the method of seeming to float the roof over the floor. His own house is typical of these styles.

■Within this 8m² rectangular concrete box life is going on. It seems as if the bare minimums of life have not been provided. The only provision is the kitchen constructed within the box. The bathroom and lavatory have been built outside of the living room and suggest that they are pieces of equipment as opposed to rooms. The storehouse surrounding the terrace like a fence and the roof that look like an umbrella have also received the same treatment.
Despite the fact that this house has been simplified down to the bare minimum, a conjuring up of delicate detail has provided good results, such as the transparent fencing and gate, the front door and entrance hall, the impression of oneness between "space A" and the verandah. The floor from inside "space B" to the outside has been left on the same level. Both wide open and clamped shut rooms. The life-styles of a young couple would perfectly match the design here. In spite of the fact that the architecture has been designed with a minimum of decorations, it is not rough or filled with urban unconventional susceptibility. This house is reminiscent of the "Sky House" built by Mr. Kiyonori Kikutake, who was Mr. Ito's teacher. The epoch-making housing that appeared in Japan after the war always seems to be based within similar types of simple architecture.

1

1 東側からの全景。幹線道路から入った小路の突き当たり右手にあるが、建て込んだ住宅街のなかで驚くほど開かれた家。
2 北東側より見る。エントランス・ホールは、ほとんど外部であるとはいえ、不思議な空間性を帯びている。1階「スペースB」へのドアが見える。

1 A complete view from the east. It is situated on the top right of the small alley which in turn is connected to the highway. However, considering the fact that it is built in an area packed tight with houses, it is surprisingly well developed.
2 A view from the northeast. Despite the entrance hall being situated practically outside of the building, it has a strange sense of being larger than it really is. The door to "space B" on the first floor is in view.

2

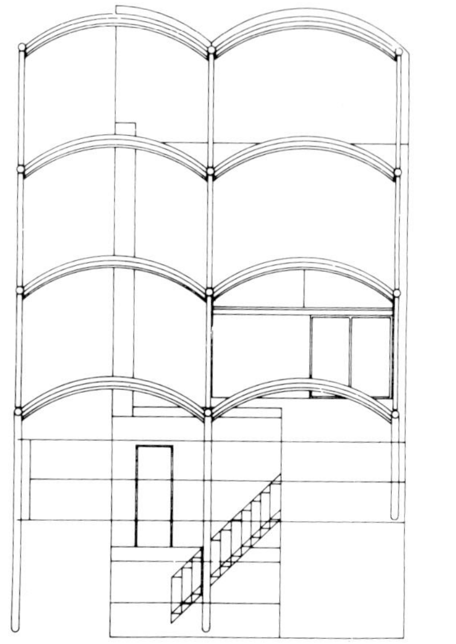

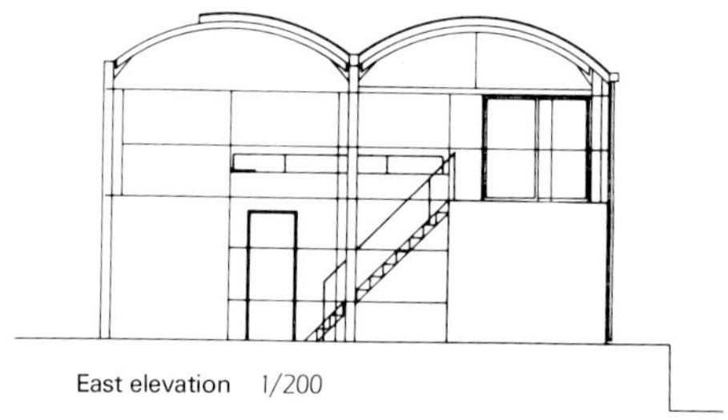

East elevation 1/200

3

4

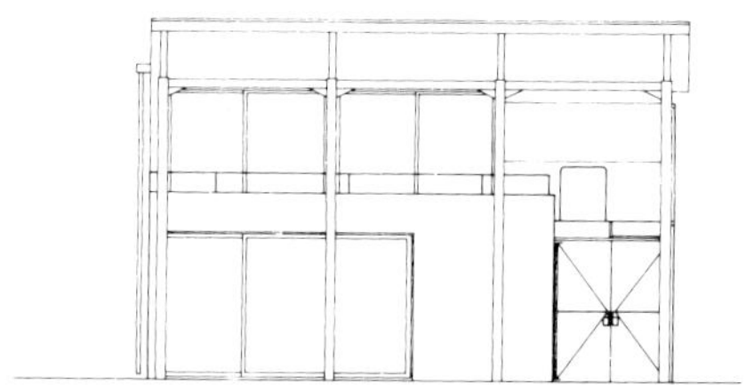

South elevation

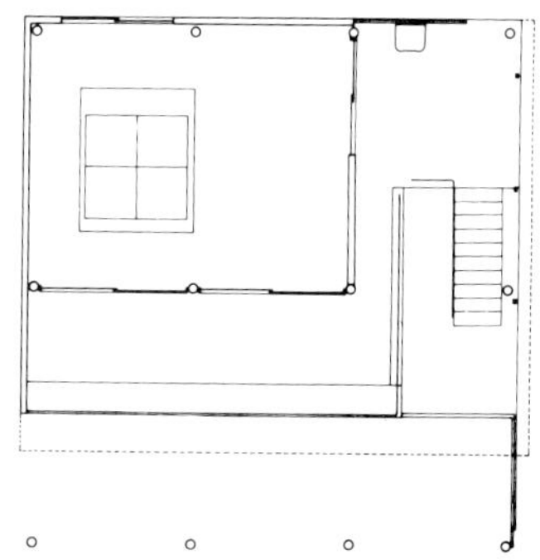

Second floor plan

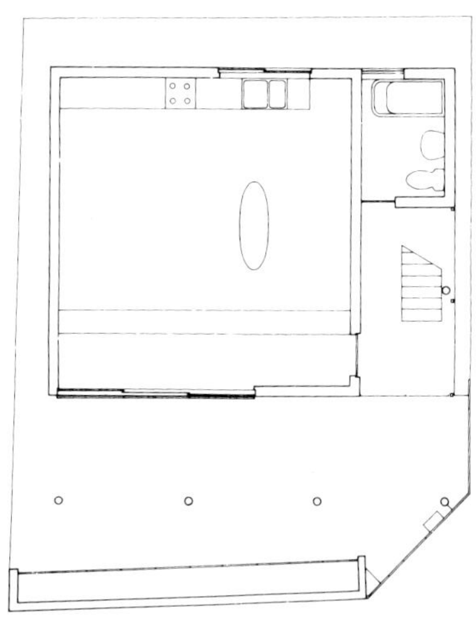

First floor plan 1/200

3東南側。倉庫がそのまま塀のように敷地いっぱいに延びて足元を隠している。
45ヴォールト屋根はH鋼の梁にデッキ・プレート、カラー鉄板平葺き。養生枠のネットはほとんど透明に近い。
6ネットとコンクリートとの対照。玄関ドア(これもネット)の内外で、同質の素材が微妙に違う表情をもつ。

3The southeast side. The storehouse has been stretched in size to conceal the base.
45The vaulted roof is made of I-beams, deck-plates and colored sheeting. The netting is almost transparent.
6The contrast of the netting and concrete. The inside and outside of the front door (also netting). The same materials have a slightly different appearance.

5

6

7

7スペースB。調理と食事とリビングに当たる一室空間。床は2段上がって入口へ、さらに外部のテラスに連続する。外側の透明な印象と対照的な、堅固に閉じ、慎重に開口部をとったコンクリートの箱。
8 2階のスペースA。寝室になっている。合板捨張りの上ラミン仕上げ。ヴェランダはコンクリート金ゴテ仕上げ。
9入口側からスペースBを見る。床は同じコンクリートのテラスに続く。キッチンの収納扉はパンチング・メタル。土間のコンクリートは金ゴテ仕上げ。10cmずつの段差。

7Space B. The area designed for cooking, dining and living. The level of the floor ascends two steps and then continues out into the terrace. The wide open and tightly colsed concrete box attains this impression from the contrast of the image of transparency.
8The second floor space A. This is the bedroom. The upper level is laminated, and the verandah has been finished by a trowel.
9A view of space B from the entrance. The floor continues onto the concrete terrace. The cupboard doors in the kitchen are made of pounded steel. The concrete floor which was finished by a trowel has been raised by 10cm.

8

9

10

11

12

13

14

10 門の郵便受け、玄関ドアの錠、照明器具、階段——すべてのエレメントが砂漠のなかに点在しているように、くっきりと意識される。
11 12 部屋の内と外のコントラスト。
13 2階、階段を上がった所に洗濯槽。その真下に浴室・便所。建築家の思いきった提案をよく使いこなしている。
14 小路の突き当たりに見えてくる北側からの全景。南側には木立が残っている。

10 The appearance that the letter-box, key-hole and light fittings on the staircase are elements scattered in a desert is prominent here.
11 12 The contrast of inside and outside.
13 The washing area at the top of the stairs. The bathroom and lavatory are directly underneath. The suggested image of the architect has been well utilized.
14 A complete view of the whole area from the north. The thicket that still clings to life on the southeastern side.

石井和紘[ISHII Kazuhiro]

反ブルネレスキ・ハウス[Anti-Brunelleschi House]

1

■日本の建築家のほとんどが、素材の自然な質感を大切にしている。木なら木らしく、鉄なら鉄らしくあることに意を注ぐ。それらの素材が単なる形態に抽象化されることや、別の被覆材で隠されてしまうことに対して警戒しつつ設計を進める傾向が強い。

そのような素材の真実性への依存に対して、石井和紘は独特なスタンスを保っている。それは、木や鉄や、また打放しコンクリートを、積極的に被覆したり、反構造学的な形態をあえてつくり出すわけでは、必ずしもない。彼は、木も鉄もコンクリートも、そのままに露にしていることが少なくないが、素材としてどこか不安定な部分がむき出しになっているような印象を受けるのである。

石井の手掛ける建築は、つねに歴史的技法が過剰に参照されている。それが、彼の仕事が単なるパロディであるようにも、他に例がないほど生真面目であるようにも見せている傾向があるが、いずれにしてもその結果、例えば打放しコンクリートは、さらにその現実的な生の表面を剥がされてしまう。素材が生活に与えようとする温もりを奪われるのである。それはいいかえれば、近代建築が成熟させてきた、素材の反形式性、つまりそれだけで成り立つ現実性を否定することの印象であり、不安感なのである。

■2本1組のダブル柱というヴォキャブラリーひとつで、住宅のエレヴェーション全面をつくりあげてしまった。道路の外観も中庭側の外観も、同じダブル柱で1、2階とも解決し、屋上の手摺にまでその痕跡を残す骨組(スケルトン)で一貫している。ダブル柱を採用すると、出隅入隅の各コーナー部のディテールも無理なく解決することができるという石井の基本的な考えがあり、この着想を建築史上の諸例(ルネサンスから現代まで)に参照させていることが、この住宅の名称に窺われる。

南側から見れば、左右のウィングを前方に突き出したシンメトカルな邸宅風の構えは、徹底した全面ガラス張りの現代的な表情の底に、古典的様式の暗さのようなものを感じさせる。あるいは様式の廃墟ともいうべき印象。それは、打放しコンクリートの骨組にも影響して、石の神経を剥き出しにしたとでもいうべき奇妙な味わいをかもし出している。

老夫婦+夫婦+子供2人、2世帯で3世代が住む。東のウィングに若い夫婦の家族が、西のウィングにその両親が住む。温室のような建物だが、必要な部分には廊下とその奥を間仕切って、書斎や洗面所などを確保している。広い中庭を挟んで、親子のガラス張りの部屋が向き合う。日常の光景には、不思議な明るさがある。

2

3

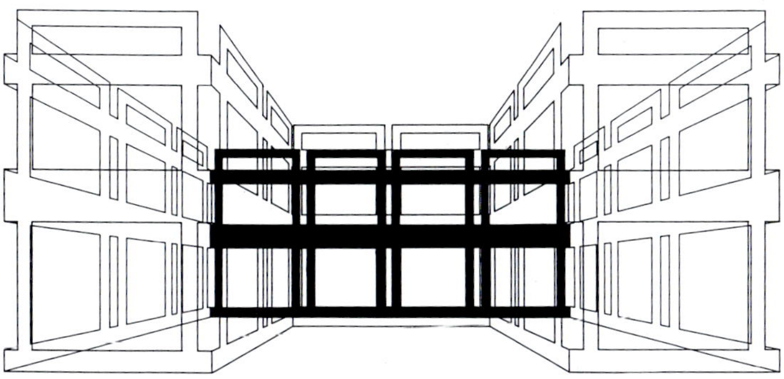

■1屋上庭園。建物のダブル柱がそのまま手摺りとなって顔を出しているので、この住宅の基本的な考えがよくわかる。
■2中庭側。建物の外観と内側。
■3食堂より廊下と中庭を見る。中央にコーナー部のダブル柱。

■1The roof garden. As the double-pillar reveals the transfiguration of the railings, the basic concept of the house can be easily undertstood.
■2The courtyard. The outside and the inside of the building.
■3A view of the passage and courtyard from the dining room. A double pillar in the central corner.

4

5

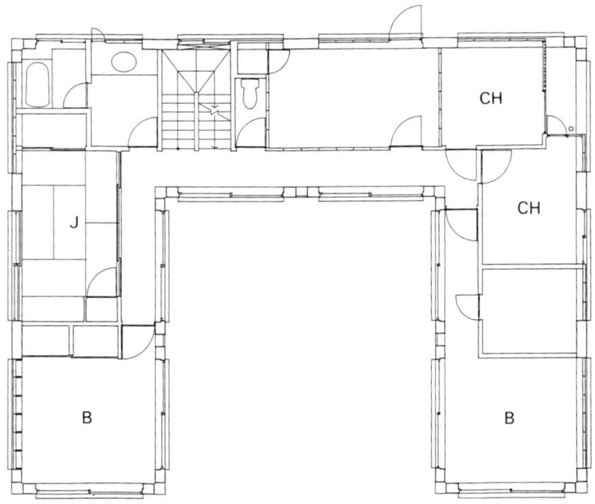

Second floor plan

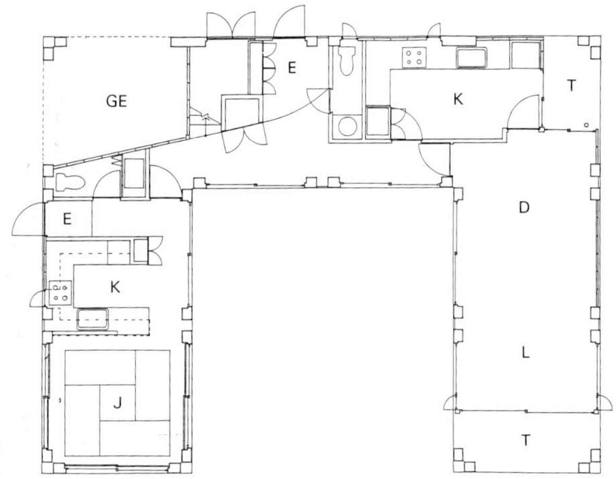

1/200
First floor plan

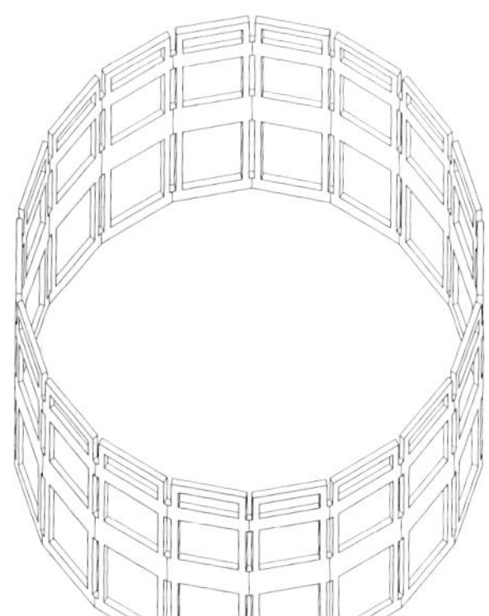

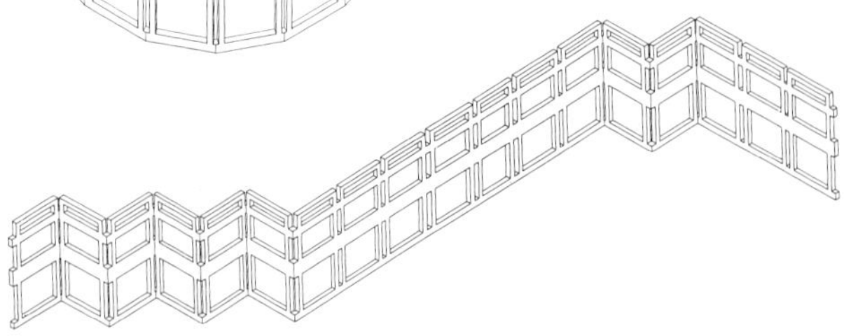

■Most Japanese architects set high value on the use of natural materials. They treat wood and iron in their original forms. They tend to put special effort into setting forth the architectural design while taking precautions against these materials being abstractly altered to simple formations or hidden by the use of other materials.

Kazuhiro Ishii maintains a unique stance towards the dependence of the reality of such materials. It is not that he always covers up wood, iron or ferroconcrete to create anti-structural formations. His architecture merely exposes the texture of the materials as they are and gives the impression that they have been deliberately exposed to show off the unstable parts within them.

Mr. Ishii's technique seems to learn towards the history of architecture. His works seem to give off the impression that they have either been built as parodies, or with exceptional earnestness. It is almost as if the ferroconcrete has been torn away from itself in order to reveal its crude inner surface. The warmth that the materials try to provide to life is seemingly stolen. In other words, it is the antiformation of materials that modern architecture has been working on to impress upon all that the materials cannot exist as they are. Otherwise they are doomed to being non-entities.

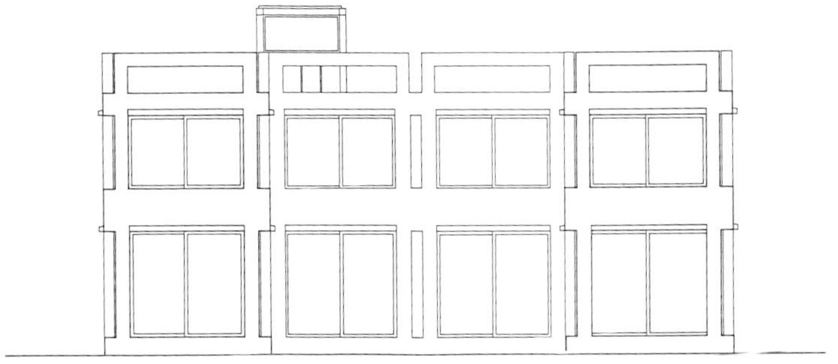

Elevation 1/200

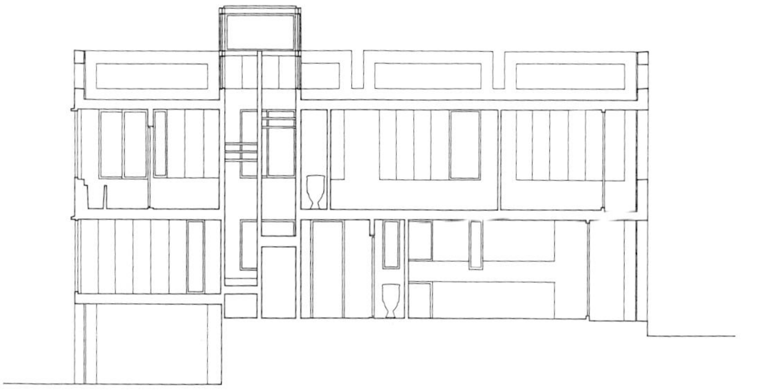

Section 1/200

■This elevation has been achieved by a double-pillar. This double-pillar method has been used for the first and second floors facing the street and courtyard, and even up to the railings on the roof which leaves a sort of skeleton tracing. Mr. Ishii has worked on the basis that the detail of each corner can be solved by the adoption of the double pillar, and history shows that this is not a new conception as it has been used from the times of the Renaissance.
Looking at the southern side, the symmetry of the condominium-style structure which projects into both wings gives off a typical feeling of gloom as is mirrored from the classical times, but this has been contrasted with the use of modern glass. It can also be imagined as a pile of old ruins. This style has even managed to exert its influence upon the structure of the ferroconcrete to produce a peculiar aroma as if the nerve center has been exposed.
This construction has been separated into two households and is occupied by a three-generation family of grandparents, their children and grandchildren. The east wing is occupied by the young couple and their children, and the south wing by their parents. Although the building is like a greenhouse, the study and bathroom have been created by dividing the passage and the space beyond. The glass panelled room which overlooks the courtyard is shared by the parents and their children and adds a magical touch of brightness into daily life-styles.

4南西側から見る。ダブル柱は横力に対してつながれている。
5屋上から中庭を見下ろす。
6中庭側のコーナー部。直角に位置するダブル柱。

4A view from the southwest. The double-pillar compounds the power of the walls.
5Looking down on the courtyard from the roof.
6One corner of the courtyard. The double-pillar stands at a 90° angle.

6

7

8

7食堂。中庭を挟んだ向かいの部屋が、親の厨房。
8中庭側コーナー部を内側から見る。
9 2階廊下。左手前に屋上への階段。

7The dining room. The room on the other side of the courtyard is the old couple's kitchen.
8A view of one of the courtyard's corners from inside the house.
9The passage on the second floor. The stairs on the left lead up to the roof.

9

10

11

12

10階段室。親のゾーンと子のゾーンの境界ともなっている。
11階段室。ペントハウスを見上げる。
12 2階和室。つながれたダブル柱。
13中庭のファサード。コンクリートのフレームとその内側の生活装置。

10The stairwell. It also acts as a boundary between the territories of the old and young couples.
11Staircase, looking down from the penthouse.
12The Japanese room on the second floor. The jointed double-pillar.
13The facade in the inner garden. A concrete frame and its inside.

13

石井 修 [ISHII Osamu]

目神山の家8 [House in Mekamiyama 8]

■石井修のもっとも高名な建築の仕事は、西宮市の目神山と呼ばれる丘陵の住宅地の、わずか200mほどの道に沿って建つ9軒の住宅だろう。50歳半ばにして、ここに自邸をつくったのを皮切りに、現在までの十数年に次々に依頼を受けて設計してきた結果、日本では例のない統一的な住宅地景観が成熟してきている。

その住宅群は、緑や石や大地のなかに溶け込むように、いわば隠されている。しかしそれは建築が、庵のように素朴になることではない。石の多い目神山の斜面のなかに家を配すためには、部屋々々を分散し、それぞれを長い通路で結ぶというプランニングになることが多い。それは道から玄関までの長いアプローチの延長でもある。これらの通路は自然のままに放置されることなく、地を割り、宙を飛ぶブリッジとなって部屋々々を連結する強い架構体として見え隠れする。ここにコンクリートや石組みという、男性的な素材や構造性や表現が呼び込まれる。コンクリートを基壇として、その上に乗せられた木造の家の素朴な喜びを感じさせる表情。外的条件から屋根までが大地の延長として扱われることもある。土と緑で蔽われた勾配屋根や段状屋上庭園。地下に潜む意志と地表を崩して姿を見せる意志。

■目神山の10軒近くの住宅群のなかでもこの住宅は異色である。それは第一に、敷地の特異な形状によっている。6m幅の前面道路より狭い平均4m強の奥行、間口は約70mという細長さであり、しかも背部に迫る公有の自然林との境界へ、3mほど低くなった傾斜地でもある。

この敷地のほぼ半分を使って建てた住宅は、地上階では一番奥行きのある北西端に主寝室を置いただけで、あとは通路とギャラリーにし、地階に、北西端に居間・食堂、反対側に和室を置き、その間を結ぶ長い廊下に個室やその他の小部屋をとり付けている。家族5人が利用する週末住宅であり、将来は成人した子供の新居として利用することを想定している。

この住宅の特異性は、敷地に規制されつつ、その制約を建築家がむしろ楽しんで積極的にデザインしている点にある。その結果、目神山の他の住宅に見られる、木の太い柱や壁を多用した、どちらかといえば素朴で隠棲的な家の表情に比べて若々しく都会風な雰囲気を発揮している。石井の表現の幅の広さに驚かされると同時に、その好奇心の旺盛さは石井の仕事の脈絡を分かりにくくしているといえるかもしれない。

道路のファサードは、低い石組みの擁壁の上にさらにいっそう立ち上がった長い壁面しか目に入らない。その屋上にも擁壁の上にも植栽が、家を蔽い包みつつある。

■Probably Mr. Ishii's most famous work so far is the row of nine houses be built along a road within the limited length of only 200m. It is located in a residential area on a hill called Mekamiyama in Nishinomiya. He has been designing house after house for over ten years now, ever since he designed his own house in his mid-fifties. As a result, he has developed a combined housing style that has no precedent in Japan.

This group of houses has literally been concealed in the greenery, stones and earth; almost as if they are being absorbed by Mother Nature. But this does not mean that his architecture is simple. Quite often he has designed the rooms to be scattered along a passage that is in fact an extention of the long approach to the front door from the road. These passages then disappear and reappear by connecting each room like a strong bridge. He likes to use a very masculine approach for the materials, structure and expression, and tends towards concrete and stone. That is the impression that a wooden house built onto a concrete base gives off. Even the roofs are often treated as an extention of the ground depending on external conditions. The slanting roof and stepped roof-garden are surrounded by greenery. These constructions have the combination of intentions of concealing the basement but opening the roof up to the sky.

■This house is a unique example of the row of houses in Mekamiyama. Part of this is owing to the shape of the site. It has a length of 70m and a width of only 4m, which is narrower than the road. In addition to this, the land is sloping and 3m lower than the boundary of the public woods that spread out at the back.

This house is built on approximately half of the available site. Only one room occupies the top part of the house and that is the main bedroom which is the largest room in the house. The gallery and passage takes up some of the space, however. The basement consists of the living room, the dining room and the Japanese room, and the long passage runs straight through it with two more rooms. This is a weekend house for a family of five, and it could even be used as a house for one of the children when they grow up. Another part of the uniqueness here is that it is obvious that the architect had a great deal of fun designing a house within such a limited area.

As a result, a wealth of expression has been created here by Ishii, and the urbanal atmosphere of the house is difficult to find elsewhere within Mekamiyama which tends to attract a lot of retired people. From the road, the only part of the facade that can be seen is the upper wall that stands higher than the low stone retaining wall. The plants on the roof and retaining wall are slowly confining the house.

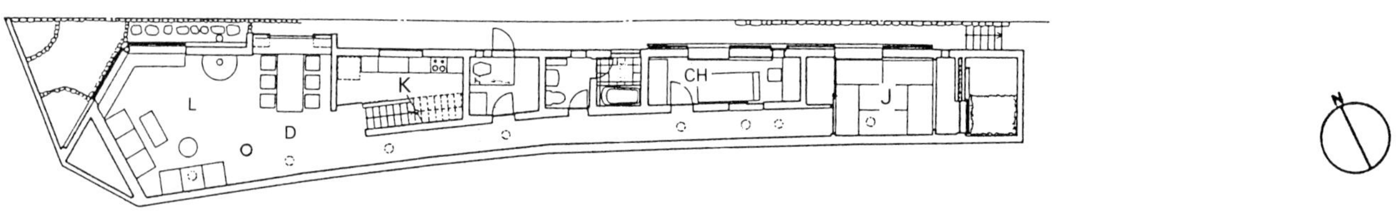

Basement floor plan

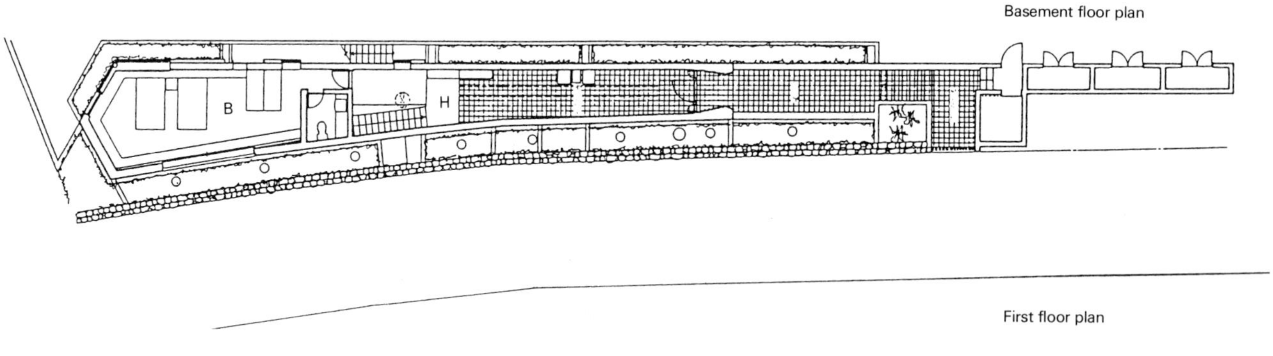

First floor plan

1/300

1

2

3

■1道路側からの全景。手前の家も、石井修の設計。
■2門とアプローチ。門を入った左手に玄関ポーチへの階段がある。
■3玄関ポーチへの階段。左右の擁壁に植込みがある。石井の建築には珍しい白いタイル。

■1A complete view from the road. The house by the side was also designed by Mr. Ishii.
■2The gate and the approach. There is a flight of steps leading up to the porch on the left-hand side of the gate as one enters.
■3The steps up to the porch. There is shrubbery on the both sides of the retaining wall. The white tiles that feature only occasionally in Mr. Ishii's architecture.

4東側より地下階を見る。上部に玄関部分が見える。
5住宅の東側。上部に寝室、下に居間。
6地階。居間より通路を通して見る。打放しコンクリートとレンガ積みの壁。

4A view of the basement from the east. A part of the entrance can be seen at the top.
5The eastern side of the house. The bedroom in the upper part of the building and the living room in the basement.
6The basement. A view from the living room through to the road. The wall made of ferroconcrete and bricks.

7

8

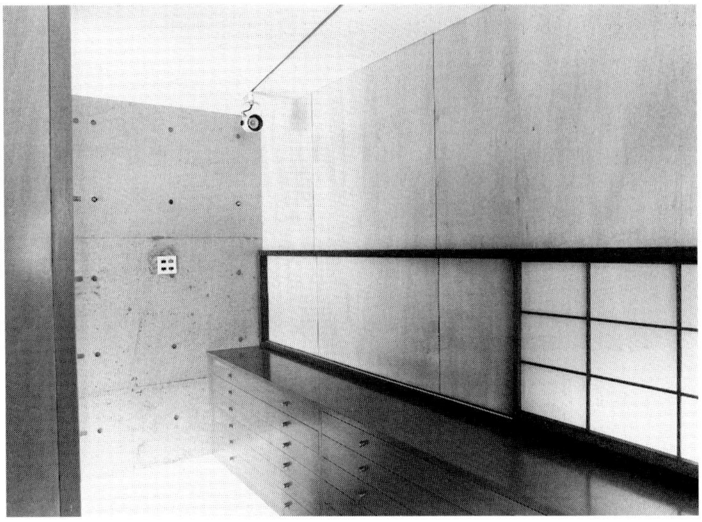

9

7地階。居間の大きな窓に公有自然林の境界線が迫っている。
8 1階ギャラリー。ナイフのように突き出た階段の手摺。飾り棚のような使い方。奥に主寝室のドア。
9 1階。主寝室のコーナー。

7The basement. The boundary of the public woods approaches the living room's large window.
8The gallery on the first floor. The banisters of the staircase that projects like a knife. It is used as an ornamental shelf. The door to the main bedroom is at the back.
9The first floor. A corner of the main bedroom.

10

10玄関ホールがそのまま美術館を飾るギャラリーになっている。天井は、音響の効果のため、フィルムケースを打ち込んだコンクリートのまま。
11 1階。主寝室。腰壁がそのまま収納になっている。和風で統一している。

12地階。居間と食堂。洋風の趣向。

10The entrance serves the same purpose as a museum gallery. The ceiling is tamped concrete embedded with film cases to promote acoustic effects.
11The first floor. The main bedroom. The panel is in fact a cupboard. It has been decorated in Japanese style.

12The basement. The living room and dining room. A western-style has been chosen here.

11

12

光明台の家 [House in Komyodai]

■傾斜地に建つ目神山の家々が、個々に分離した部屋とその間を結ぶ通路や内庭といったプランニングになる場合が多いのに似て、この住宅もコンクリートで囲われた箱状の5つの部屋——アトリエ、台所、和室、浴室・便所、洗濯室・納戸を、分離・独立させた形で敷地のなかに配し、これらの箱を結んだ中間の領域を居間や食堂にしている。
5つのコンクリートの箱は、木造の2階と屋根裏をつくる、いわば基礎柱にもなっている。さらにその屋上はすべて土をのせた庭となり、花や野菜やハーブの栽培に利用されている。同時に地上にも、4つの変化に富んだ庭をつくり出しているし、道路側には、パーキング・スペース、アクセスが確保されている。合掌造りの大屋根の下からウィング状に突き出た5つの箱が、敷地を分節しているからだ。
目神山の家に似て、点在する部屋々々を基点とする構成は、この家が建つ泉北ニュータウンの分譲地、すなわち思い思いに意匠を凝らした住宅が並ぶ典型的なベッドタウンの区画地のなかで、全く別のスケールと表情を帯びて出現する。急勾配の屋根を支える太い双対の柱、左右に水平に伸びるシンメトリカルな完結性が、ファサードの主調になっている。

■This house has many similar characteristics of the Mekamiyama houses, built on a slope, in the sense that the rooms are scattered and individual, and they are connected with a passage and courtyard. It has five concrete box rooms — an atelier, the kitchen, the Japanese room, the bathroom that is a combined lavatory, and the scullery with storage space. The combined living and dining room make up the inter mediate space for the other rooms.
The five rooms are the equivalent of the basic pillars that go to make up a two-story wooden house with attic. Additionally, the roof-terrace can be used for cultivating flowers, vegetables and harb. Simultaneously, the four gardens on the ground produce external areas for parking or access. The five rooms projected beneath a large slanted roof separate the site.
The construction on the scattered rooms is very similar to the houses in Mekamiyama and appears with a quite different scale and expression in Izumikita new town. There is typical of areas where the designs of houses all differ from one another. The double-thick pillars support the steeply slanting roof, and the completed effect is that of symmetry which stretches over the whole construction and acts as a keynote in the facade.

1

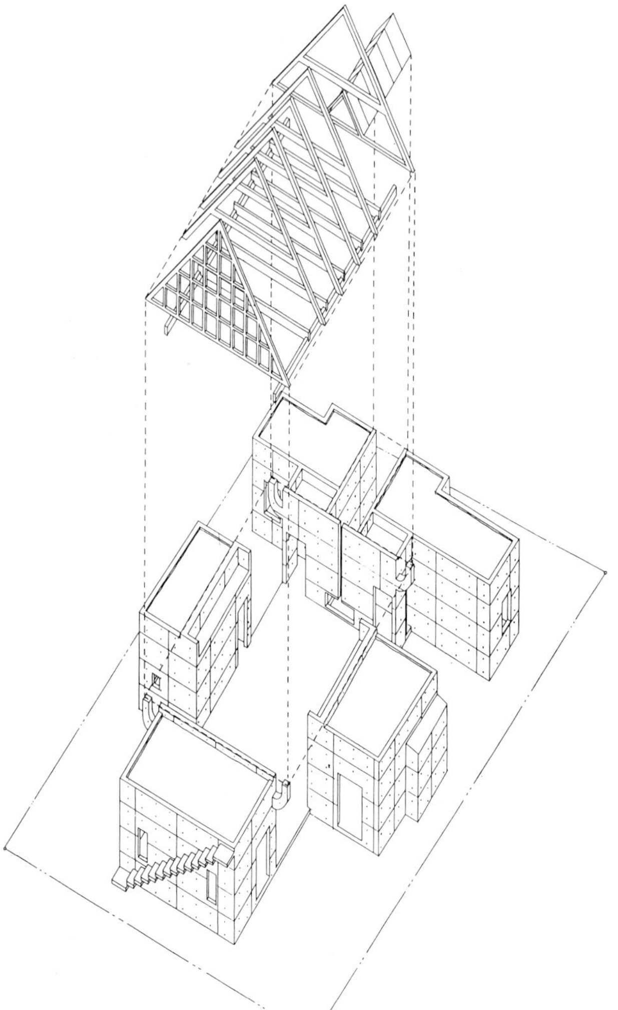

2

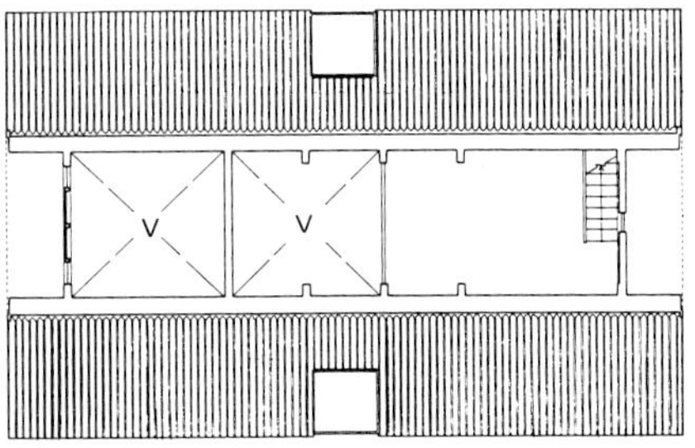

Rooftop plan

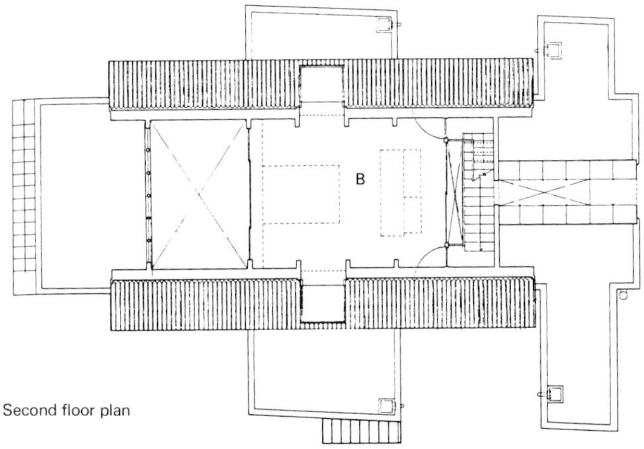

Second floor plan

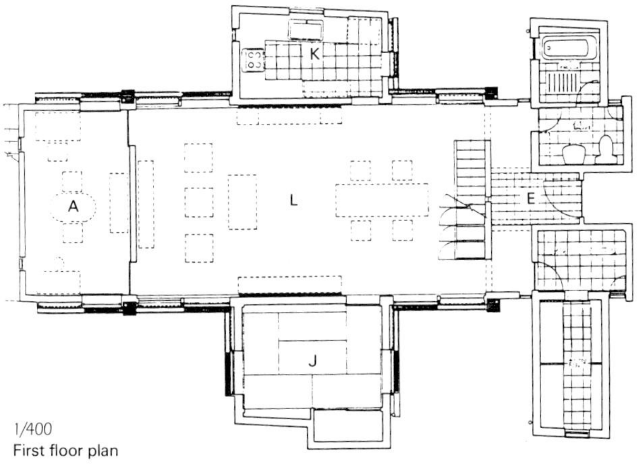

1/400
First floor plan

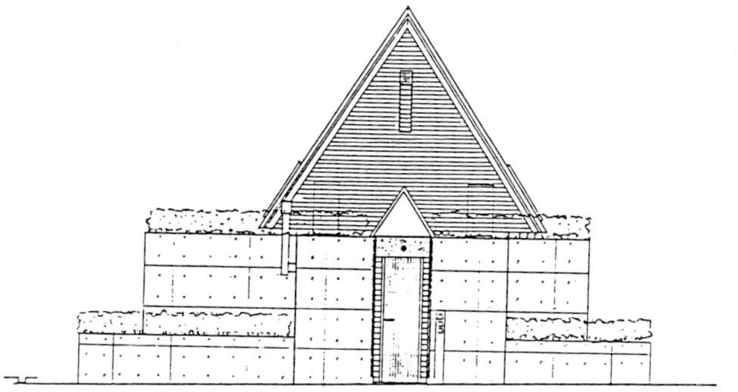

North elevation

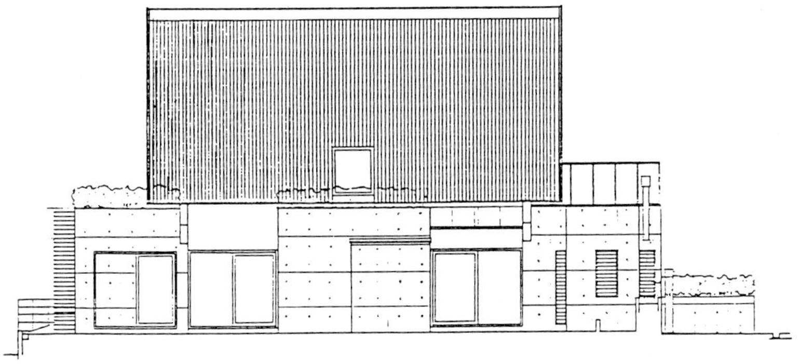

East elevation

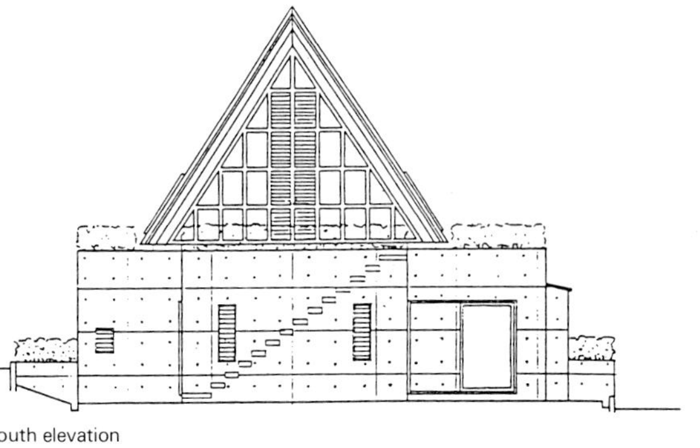

South elevation

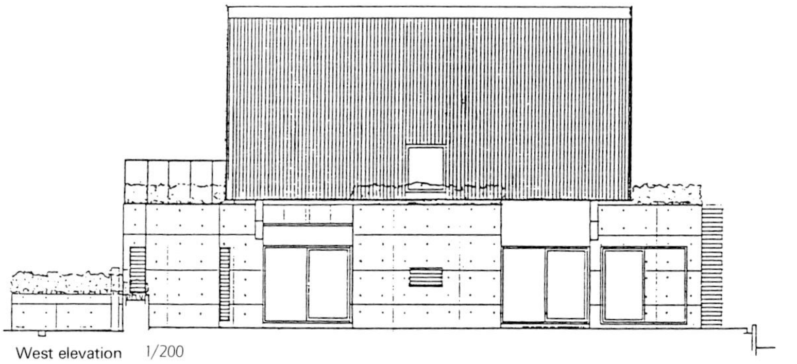

West elevation 1/200

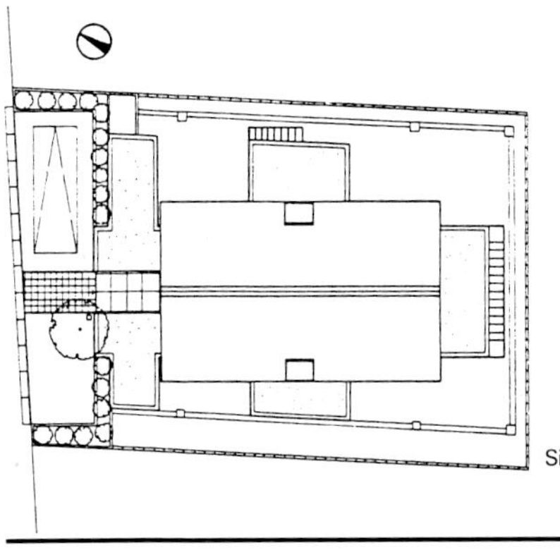

Site plan 1/400

3

1 正面。門構えのような双対のコンクリート柱が、実は便所と洗濯室。縦長の開口部を目立たないように扱っている。
2 ガラスのキャノピーがそのまま玄関ホールのスカイライトに続いている。チャペルのような素晴らしいスケール。
3 玄関まわり。

1 The front. The double concrete pillars that resemble a gate is in fact the toilet and scullery. It prevents the feeling of height from being too outstanding
2 The glass canopy continues onto the skylight in the entrance hall. Splendid chapel-like scale.
3 The area to the entrance.

4

5

4玄関扉両脇とホールの中心にアクセントをつけるガラス・ブロック。効果的な採光装置でもある。
5居間上部の吹き抜けと開口部の表情。
6居間より食堂を見る。階段室の奥に玄関ホールとの境に立つコンクリートの壁。
7食堂から居間方向を見る。コンクリート壁の奥にアトリエ。

4The glass blocks provide the accent to the eastern side and the center of the hall. The effective lighting.
5The high ceiling reaches up to the roof and opens up the living room.
6A view of the living room from the dining room. The concrete wall stands at the boundary of the stairwell and entrance hall.
7A view of the living room from the dining room. There is an atelier at the rear of the concrete wall.

6

7

8

9

10

11

12

89北西側(玄関側)の三角形の開口部と南東端の三角形の開口部。
10 2階から屋根裏への階段。右手に主寝室の引戸、左に玄関ホールの三角形の窓。
11居間。右手にアトリエへの入口。正面左の壁の奥に和室。ここだけ桟を白く塗った障子と、打放しコンクリートと木の大梁との対照。
12主寝室は2階フロア全面を占め、前後に吹き抜けをもつ。

89The triangle openings of the northwest side (the entrance side) and the southeast side.
10The stairs leads from the second floor to the attic. The sliding door of the main bedroom is viewed on the right-hand side. The triangle window of the entrance hall is on the left.
11The living room. The entance of the atelier is on the right hand side. The Japanese room is situated at the rear to the front left wall. The contrast in combinations of the whitely painted frame of the paper sliding door, the ferro-concrete and the wooden beam.
12The main bedroom takes up the whole second floor with the open floor in front and in the rear.

槇 文彦[MAKI Fumihiko]

自邸[Maki's House]

■本書に登場する、安藤、長谷川、早川、伊東、六角は同じ1941年生まれである。日本の建築界で、ポスト・モダニズムの状況を実際の仕事に反映させ、一挙に国際的な局面に姿を現した建築家の多くがこの1年に集中した同世代であることはきわめて興味深い。
こうした1年間に凝縮された世代をさらに遡って求めるとしたら、槇文彦の生まれた1928年だろう。岡田新一、阪田誠造、林昌二、池原義郎、菊竹清訓、奥村昭雄、林雅子、今里隆等がこの年に生まれている。1950年後半から70年代前半にかけての長い、日本の現代建築においてもっとも重要な時期に、それぞれの作品の方向性を確立し、公共建築、商業建築、伝統的建築、住宅等の各分野における多様な展開を示すと同時に、ゼネコン、大設計事務所、アトリエ、大学等の互いに異なる立場のなかで、自己表現の可能性を探った、戦後最初の世代といえるだろう。
そこには、自らのコスモロジーを確立する前に、日本の現代建築が通過すべき基本的な段階——構法、コンポネント、素材、計画論等への吟味が、先行する巨匠たちの仕事の検証をも兼ねて行われた。槇文彦の柔軟さ、平易さ、明快さはそうしたハードルを幾つも越えて現れてきたことが歴然としている。

■槇邸は10年前に完成したものであり、このアンソロジーのなかでは例外的な位置を占める。槇にとっては最初の住宅建築であり、その後も純粋な戸建住宅はほとんど設計していない。しかし、日本の、とくに東京の住宅と住み方の感性を、モダンスタイルのなかにこれほどの平易さと微妙さで見せた例としてはまれであり、他の建築家も、その局面ではこの住宅を受け継ぐような仕事をしていない。先行する例があるとすれば、槇が少年の頃に訪ねて強く印象づけられたという土浦亀城の自邸だろう。40年以上前に建てられた、その初期モダニズムの美しさと脆弱性は、槇邸の外壁に反映されている。
全体は陸屋根のマッスの不規則な集合体に見えるが、その形態は、軒の深い外部空間を巧みにつくり出し、同時に南面では、中央部のクロスバーが全体を統御している。屋内は、循環的な動線と、要所々々にとられた開口部でポーラスな部屋の集合となっており、それを明確化するように、プラスター・ボードとビニール・ペイントで白一色で塗りまわしている。

■Mr. Ando, Ms. Hasegawa, Mr. Hayakawa, Mr. Ito and Mr. Rokkaku, all of whom appear in this book, were all born in 1941. It is extremely interesting to know that many of the architects who have become known by the world by the effects of post-modernization are being reflected in the work of Japanese architects of a similar generation. But if we go back in history to search for the most significant year, 1928 would be an obvious choice. Mr. Fumihiko Maki was born in this year, along with other greats like Mr. Shinichi Okada, Mr. Seizou Sakata, Mr. Shoji Hayashi, Mr. Akio Okumura, Ms. Masako Hayashi and Mr. Takashi Imazato. The most important period for all of them was the years between the late 50's and the early 70's. During this period they established the trend of Japanese architecture and each made great progress in the fields of public, commercial, traditional and housing architecture. This was the first generation that discovered the potential of self-expression coming from different backgrounds.
Before the architects were able to establish their own sense of cosmology, the basic steps of construction methods, the materials, the components and the planning methods all had to be examined in detail to serve as verifications of the leading architects of the time. It is clear that the architecture of Mr. Fumihiko Maki has risen above these obstacles to achieve a sense of expression that is mirrored in his flexibility, plainness and preciseness.

■This house was built ten years ago and was then an exception to this anthology. This is not only the first house that Mr. Maki built, but also the last time that such a genuine house has been built. This house is a rare example of pure sensitivity and harmony with life-styles, especially noticable in modern Tokyo, with its simple and delicate styles. No other architect has managed to utilize this concept against such an aspect. If there is an example of his style in the past, it is probably the house of Kameki, Tsuchiura that

1

seemed to leave a strong impression on him when he visited it as a child. The external wall of Maki's house reflects the early modernization which came into effect over forty years ago.
The whole building takes on the irregular shape of a block with "flat roof", but this formation skillfully produces deep space under the external eaves and simultaneously the center cross-bar rules the whole southern part of the building. The circular moving lines and openings have been placed strategically in order to polarize the groups of rooms within. It has been painted all in white, with white plasterboards and vinyl paint to make the clarification.

2

1道路側から見る。左手に玄関まわり。庭に面した南側のエレヴェーションが見える。
2玄関前。コンクリートのテクスチャーが美しい。

1A view from the road. The area to the left of the entrance. The southern elevation facing the garden can be seen.
2The entrance. The texture of the concrete is beautiful.

3

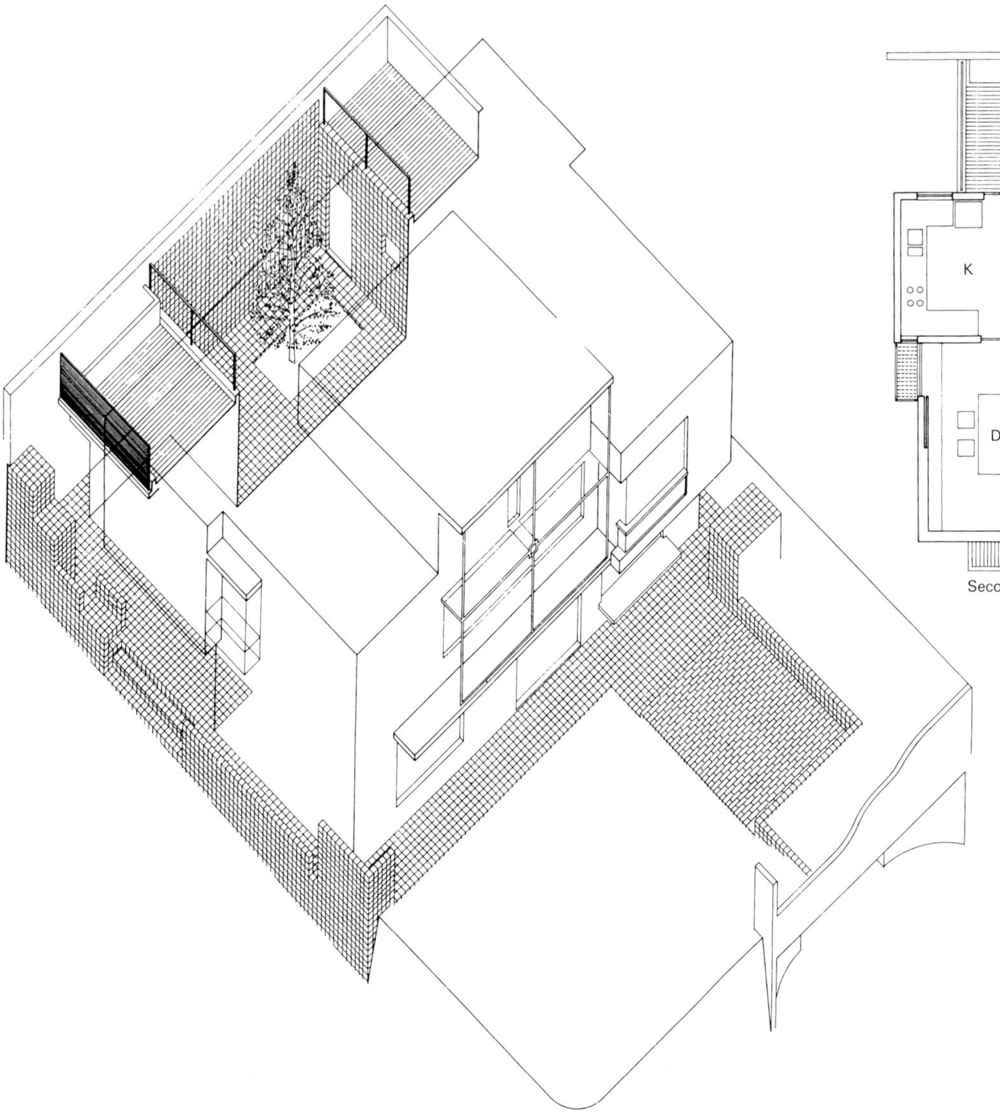

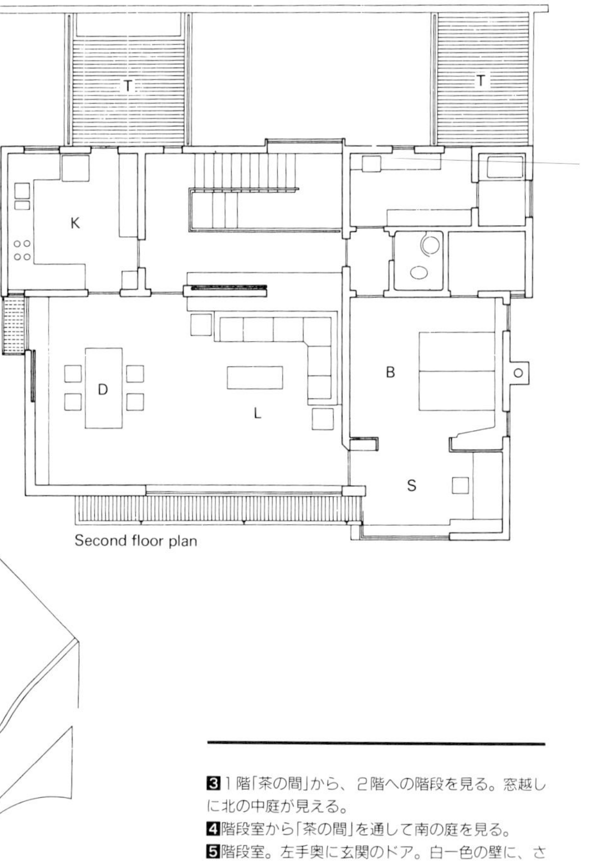

Second floor plan

3 1階「茶の間」から、2階への階段を見る。窓越しに北の中庭が見える。
4 階段室から「茶の間」を通して南の庭を見る。
5 階段室。左手奥に玄関のドア。白一色の壁に、さまざまな窓と絵画。

3 A view of the staircase on the second floor from the living room on the first floor. The courtyard in the north can be seen through the window.
4 A view of the garden to the south taken from the stairwell through the living room.
5 The stairwell. The front door is at the back to the left. The various paintings and windows on the white wall.

4

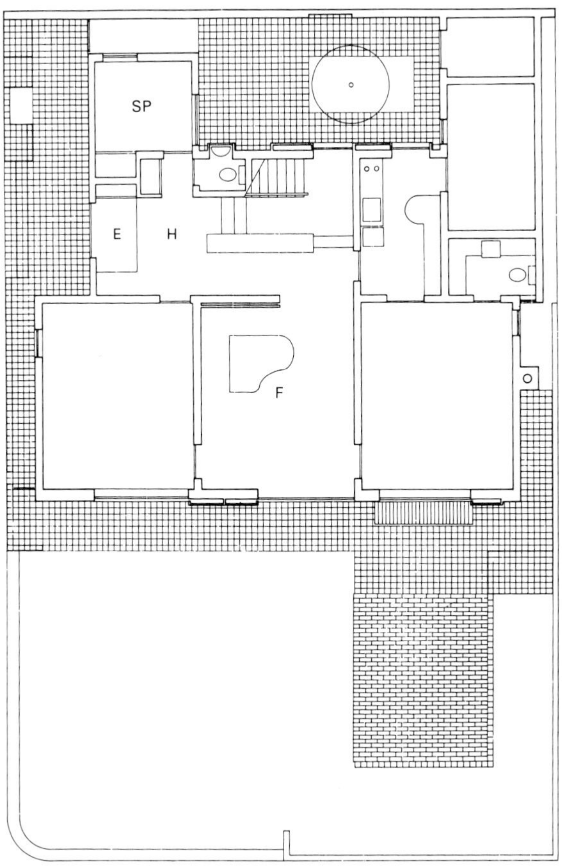

First floor plan 1/200

5

6

7

8

6南側ファサード。クロスバーが「立面の象徴」として扱われている。コンクリートと木と金属の、対照と連続。
7南側ファサードを見上げる。
8 2階階段室から居間を見る。

6The southern facade. The cross-bar is treated as a symbol for the opening. The contrast in the combinations of concrete, wood and metal.
7Looking up at the southern facade.
8A view of the living room from the stairwell on the second floor.

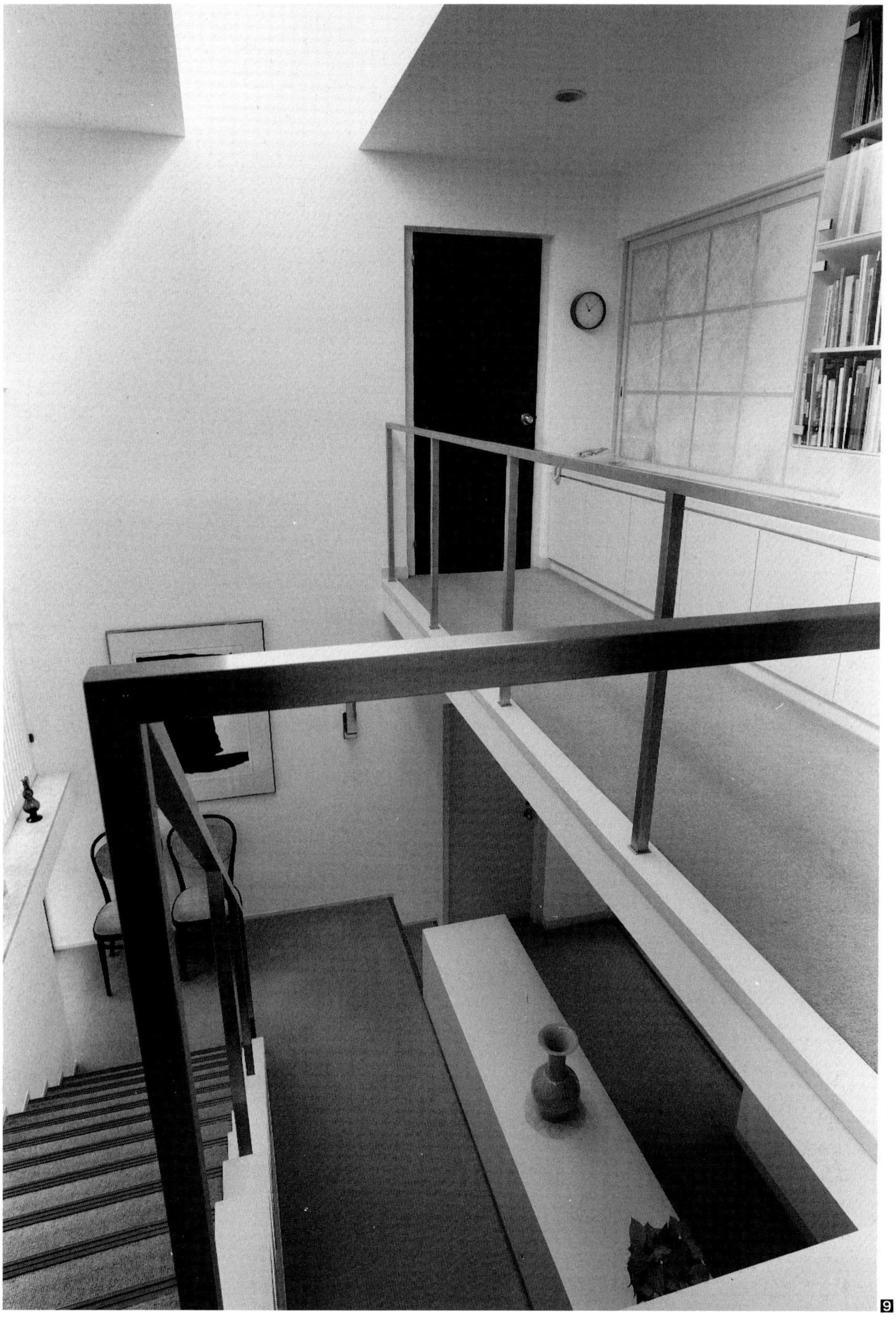

9

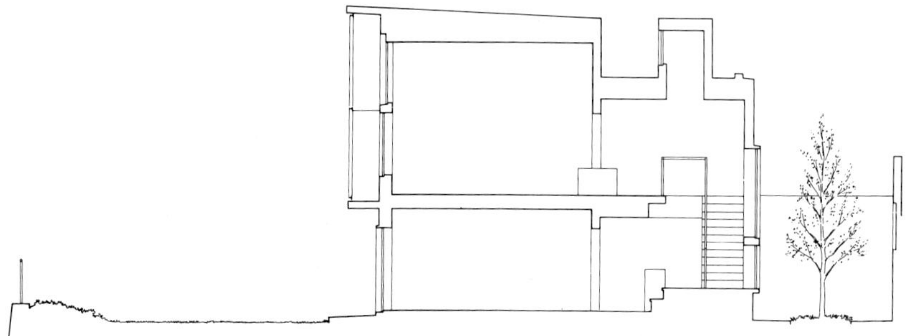

Section

9 2階廊下から階段室を見下ろす。右手障子の向うに居間、突き当たりのドアの奥に主寝室。
10 階段室から見上げる。ハイサイド・ライトが見える。

9 Looking down the stairwell from the passage on the second floor. The living room is concealed by the sliding door on the left, and the main bedroom is the room behind the door at the rear.
10 Looking up from the stairwell. The lighting fitment can be seen high up.

10

村上 徹[MURAKAMI Toru]

田方の家[House in Takata]

■広島市に在住し、建築の仕事も、市内とその周辺にあえて限っている、村上徹のコンクリートは、独自の表情をもつ。優しく、軽やかで、柔らかい。そして透明感がある。例えば、直方体の箱に寄せ棟の屋根をのせた、思いきり単純な外形の「鈴ヶ台の家」は、一室空間となっている2階の壁をプラスター・ボード白ペイント仕上げとし、天井だけ打放しコンクリートの寄せ棟の形を露出させているが、このコンクリートの肌が天幕のように柔らいで見える。あるいは「己斐の家」の大きなエントランス・コートの正面に静かに立つ、打放しコンクリートの壁は、ゆるやかに波打ち、そこから奥の第2、第3の部屋の壁にも、その波動を伝えている。演劇舞台において俳優が立ち去ったあとに残る、カーテンの揺れのような、やがて消えていく動きである。
その後に続く「黒瀬の家」や「本郷町の家」などの見事な住宅建築においても、風のわたる平原に露営する天幕のような、あるいは庭園の仮設の四阿のような、極度に単純で豪奢な印象はますます強まっている。壁性に抗らう自立的な皮膜は、その最終的な目的として、住まいの決定的な場─床を浮上させ、静謐さによって支配される空間だけを残す。それが村上の建築である。

■村上徹の住宅のなかでは、もっとも小さな敷地に建つものである。彼の住宅のほとんどが敷地いっぱいに広がり、合わせ鏡の映像のように、最小限のヴォキャブラリーを反復して、水平に透明に広がっていくのに対して、この住宅は一見すると垂直に積み上げられた印象が強い。しかし、内部はコンパクトながらも、村上の他の住宅に共通する水平的な広がりが、それぞれの階に見られる。
1階では、寝室と予備室の床をわずかに沈めて、南の小さな庭との結びつきを強めている。彼の住宅にしばしば見られる、建築の床面から湧出してくる地面あるいは眺望が、ここでも確保されている。
2階は居間・食堂が一体となり、この階でも大きく開いた南面のガラス越しに近隣の風景を見渡せると同時に、東壁面には3階の屋根の形を映した逆三角形の開口部をとっている。台所の吊り戸棚や階段と手摺も、同じ形を反復している。
3階の開口部は45度転回して、大きく東面を向き、小川ごしの林を見晴らす。明快に転換する四面のファサードは、この家が、幸福な目の高さの眺望が信じられる場所にあることを示す。光と影は、東京の都市住宅がとくにそうであるように、垂直的にも抽象的にもなっていない。

1

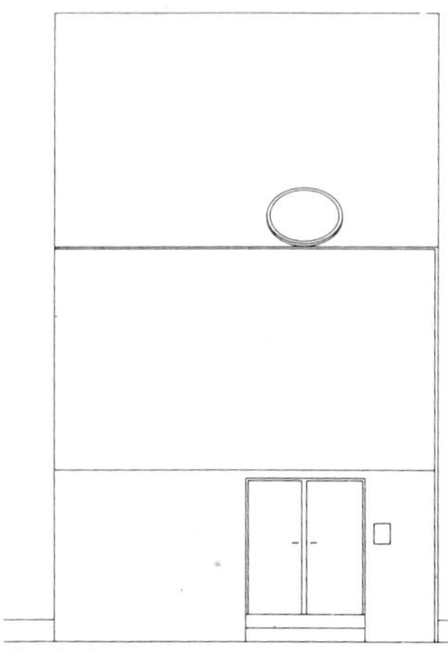
North elevation

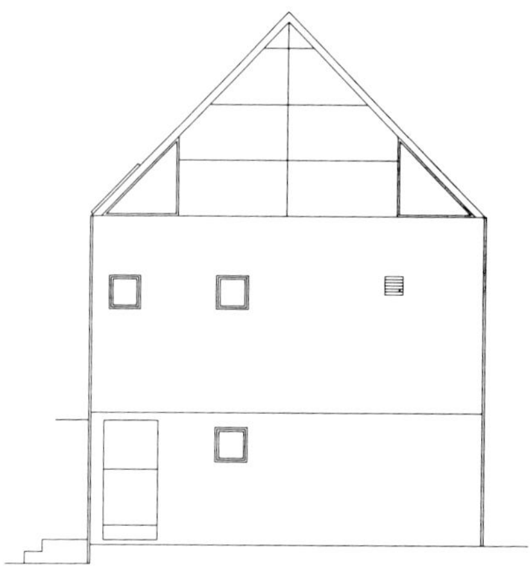
West elevation

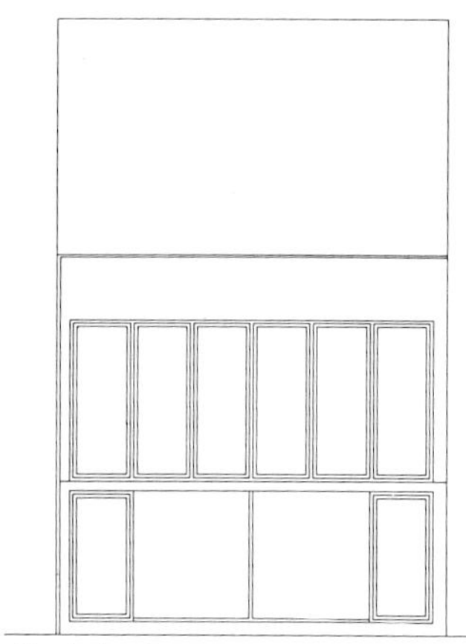
South elevation

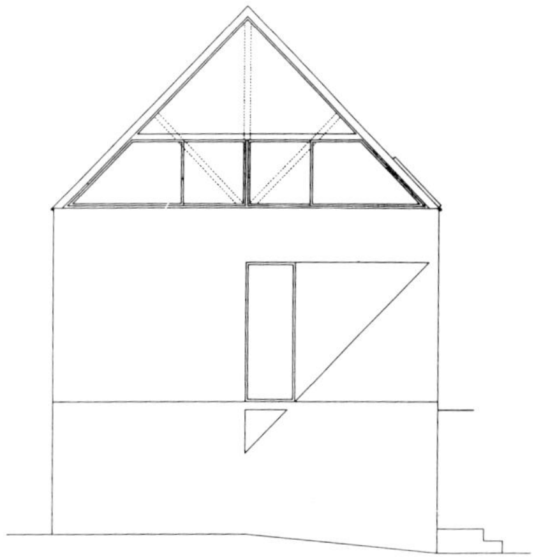
East elevation

1/150

2

■Mr. Murakami lives in Hiroshima city and most of his work has been completed within or around this city. His style of working with concrete produces unique effects.
He manages to produce a feeling of transparence, lightness, gentleness and softness. If we take a look at the "house in Suzugadai" which he designed as simply as possible, his style is evident. He has used a ridged roof on a hexahedron base, and white painted plaster-board was chosen to decorate the walls on the second floor. Despite the shape of the ferroconcrete, a ridge roof has been attached and for some reason this gives the structure a soft and gentle feel, almost like a tent.
The ferroconcrete that stands quietly on the front of the "Koi house" gently sends its waves up the wall to the second and third rooms at the rear. It is almost like the gentle movement that a theatre curtain makes having been lowered after the actors have left the stage.
This ability to create fine architectural constructions that seem like tents encamped on windy fields is also apparent in his other masterpieces like the "house in Kurose" and the "house in Hongocho". The transparent filmyness of the walls oppose the normal characteristics of a house and concentrates the energy onto the floor-space upon which life takes place, leaving an area in the total control of quietness.

■This is the smallest of houses designed by Mr. Toru Murakami. Most of his works take advantage of the full amount of space available and stretch out vertically and transparently like the images seen in double mirrors. This house, on the other hand, seems at a glance as if it has just been piled up with little consideration to space, but the interior gently reminds one of the soft Mr. Murakami touch.
The levels of the floor in the bedroom and spare room on the first floor have been slightly lowered to maintain a sense of continuation between the house and the small garden in the south. This feel of making the floor levels at one with the outside ground and lifting the scenery into the house is a common occurrence in Mr. Murakami's desings.
The living room and dining room on the second floor have been combined, and a commanding view of the neighbourhood may be seen through the south-facing window. The inverted triangle entrance, an imitation of the shape of the roof, has been located on the eastern wall of the third floor. This pattern has been repeated on the wall-mounted cabinets in the kitchen and the banisters of the stairs.
The entrance to the third floor takes a 45° turn to face the east and thereby gives a view of the forest across the stream. The clearly conversing sides of the facade indicate that this house is situated at a pleasant eye-level to make the most of the pleasant views that abound. The light and shadow are not vertical nor abstract as is often seen in houses in Tokyo.

3

4

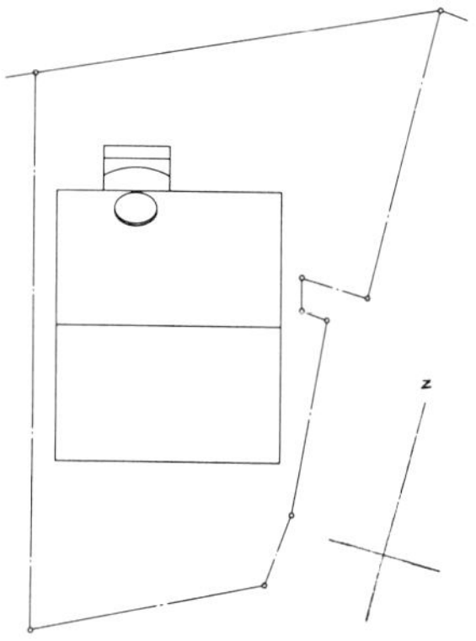
Site plan 1/300

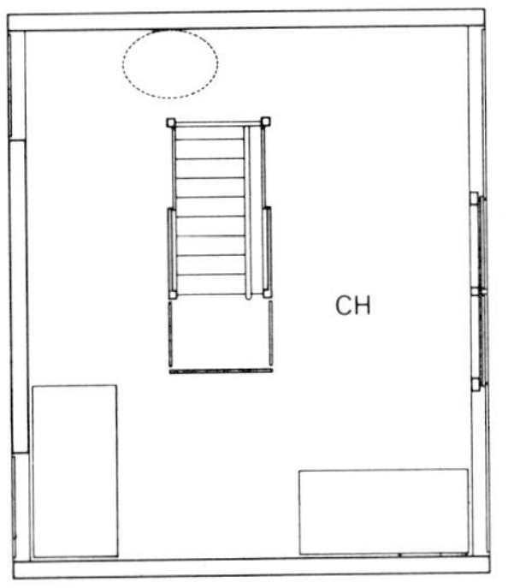

Third floor plan

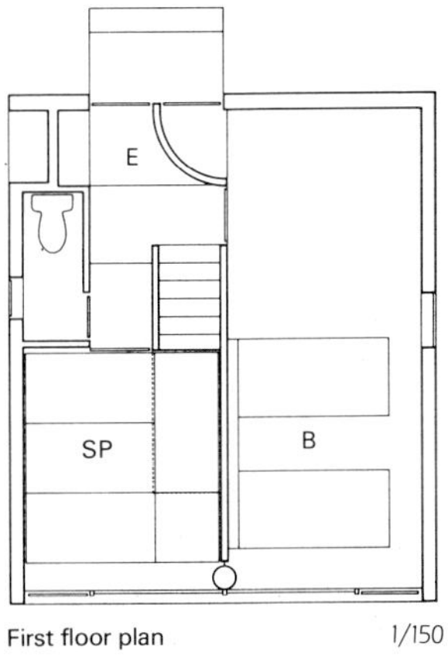

First floor plan 1/150

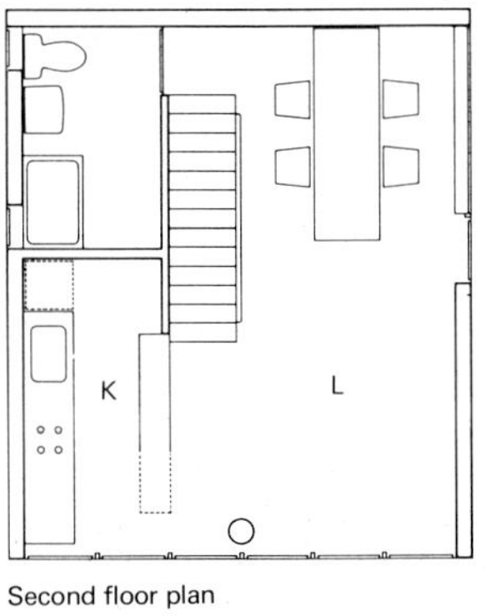

Second floor plan

1南東より見る。家の前を小川が流れ、その向うには小高い丘の雑木林が迫る。
2三角形の屋根や開口を主調とした東西ファサードに比べて、南のファサードは、全面的な開口部として見せる。
3北側、倉庫のような単純な美しさ。
4入口部分。現在の主寝室をオフィスなどに転用する場合を考えて、2つのドアを用意している。

1A view from the southeast. There is a stream in the foreground and a copse on the hill beyond.
2A comparison of the east-west facades with the triangle roof, and the north facades which are dramatically closed off with the walls.
3The northern side. A simple warehouse-like beauty.
4A part of the entrance. The two doors are purposefully fixed onto the main bedroom to account for future idea to turn it into an office.

5

5 1階主寝室の小さな三角形の開口部と2階居間の大きな三角形の開口部が、より大きな三角形を形づくる。
6 玄関口より居間を見上げる。下に、1階主寝室。
7 8 庭の高さより床をわずかに沈めた1階の予備室と主寝室。

5 The small triangle in the main bedroom and the large triangle in the living room on the second floor seem to create an even bigger triangle.
6 Looking up at the living room from the entrance. The first floor bedroom is down below.
7 8 The spare room (foreground) and the bedroom on the first floor. The floor level has been lowered to join them purposefully with the garden.

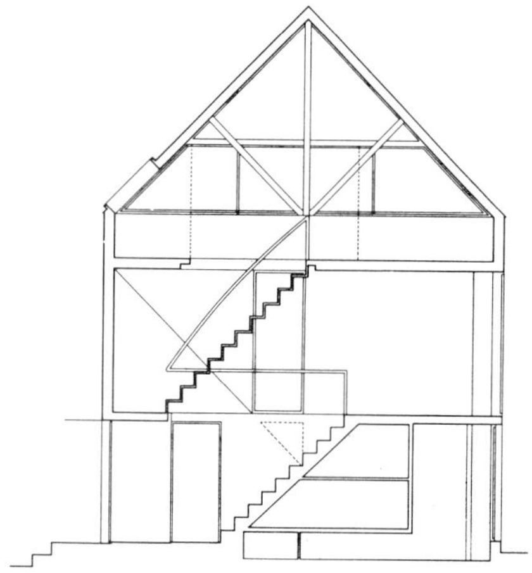

Section 1/150

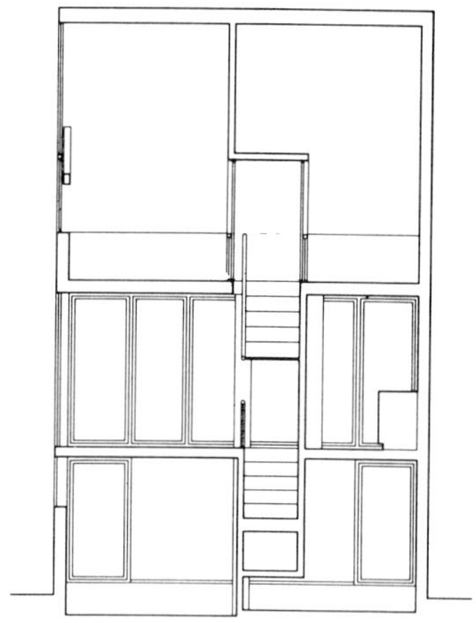

Section

6

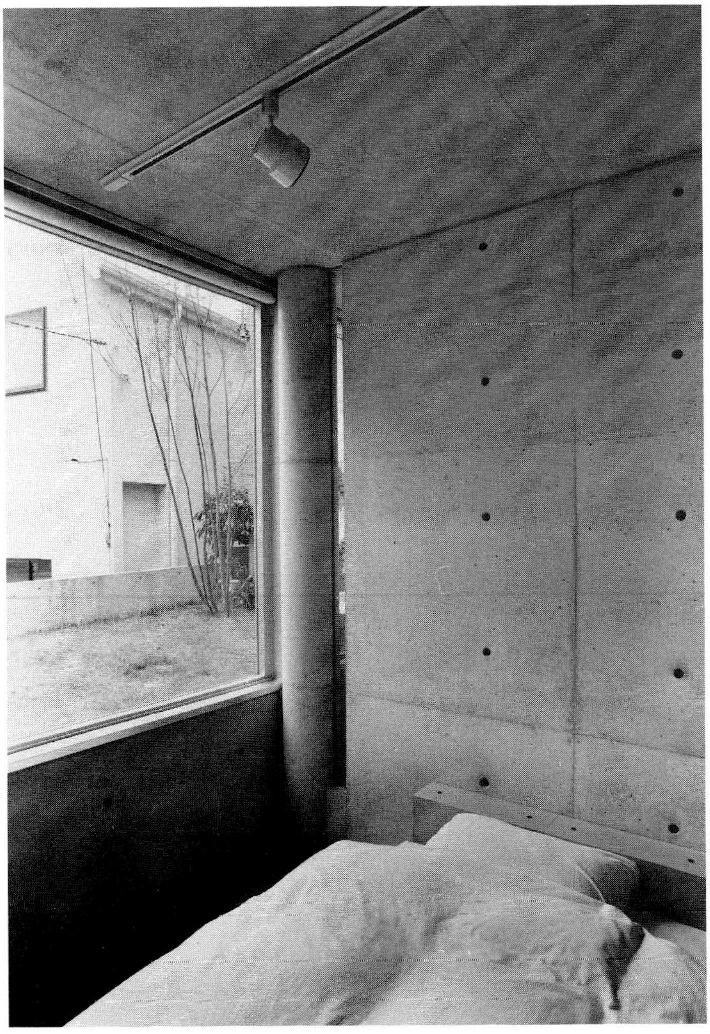

7

8

9

10

11

12

13

9居間。食堂を兼ねた一室空間。
10台所と居間を仕切る逆三角形の吊り戸棚、階段とその手摺が、東面の開口部の形を反復している。
113階子供部屋。コンクリートの腰壁の上に急勾配の屋根が乗っているが、空間は高く、広い。子供部屋はこの階段室を中心に円環状をなしている。
122階居間の南面開口部分。1階と同様、中央に円柱。
133階子供部屋から階段室を見下ろす。

9The living room. The area combined with the dining room.
10The inverted triangle cabinets in the kitchen that also function as dividers between the kitchen and living room. The stairs and banisters are also a copy of these triangle shapes that are introduced on the eastern wall.
11The children's room on the third floor. Although the ceiling is steeply sloped on top of the concrete panelled wall, the room still possesses a good deal of open space. The children's room draws a circle centered around this stairwell.
12The southwest opening in the living room on the second floor. The round pillar has been placed in the center in a similar fashion as the first floor.
13A view down the stairwell from the third floor.

室伏次郎 [MUROFUSHI Jiro]

西原の家 [House in Nishihara]

■「遺構のように」と室伏次郎は、自分の建築について、表現する。つまり、屋根も床も窓枠やドアも失くなって、ただ壁だけが立ち残っている、そんな光景を建築になかに再現したいのだ、と端的に受けとってもそう間違いではないだろう。彼の住宅には二重のコンクリート壁からなるものが多い。外側の壁は建物全体をすっぽりと包み、大きな開口部をもつ。内側の壁は、それぞれの部屋が必要とする開口部をもつが、そこに窓枠も建具もない、ただの孔である。つまり内側の壁が、気象その他の諸条件に煩わされない、生活と直結した家の「外壁」なのだ。こうしたモデルによって、室伏のすべての建築を読みとっていくことも可能である。この内側の、廃墟風の壁は、真実の壁である。同時にそれがあくまで演出されたものであり、ゲームであることも室伏は自覚している。外からは見えないように包んだり、逆に裏返しにしたりして、虚実の間を往復し、真実性のストレートな帰結、つまり家らしい家の形に帰ることを迂回している。室伏の二重の衝動、埋設への衝動と発掘への衝動が、彼の建築に考古学的な外形――半ば隠れ半ば見えている家の姿をとらせている。

■東京では高水準の住宅地、間口は狭いが南北に長い敷地に建っている。コンパクトにまとめられた長方形のコンクリートの箱だが、道路と住居との間にクッションをとることと、必要な採光・通風を充たすことの、相矛盾した要求に対応した部分が、建物の内外に微妙な表情をつけている。加えて、施主とその縁者による厳密な方位学上の判断によって、間取りが決定されるという条件、さらにファミリー・ルーム的な使い方とフォーマルな応接とが兼ねられる形式という条件などもあり、全体として、異なる住まいのヴォキャブラリーが重層している住宅である。例えば、整然と壁柱が並ぶアプローチから玄関ホール、正面の階段とアーチの天井。巨大な洋風家具や伝統的な図柄のカーペットを基調とする居間。その真上の等身大のドールハウスともいうべきロマンチックな書斎。その間を結ぶガラスのピラミッド。水を張ることのできる屋上テラス。等々が断片的に集められ、各所に埋められている。
これらは、建築、装置、家具、装飾といったおのおののレベルで自在に顔を出し、建築家も住み手も、その重ね合わせを楽しんでいるようだ。建築自身がまた、大きな一断片として住宅地のなかにある。

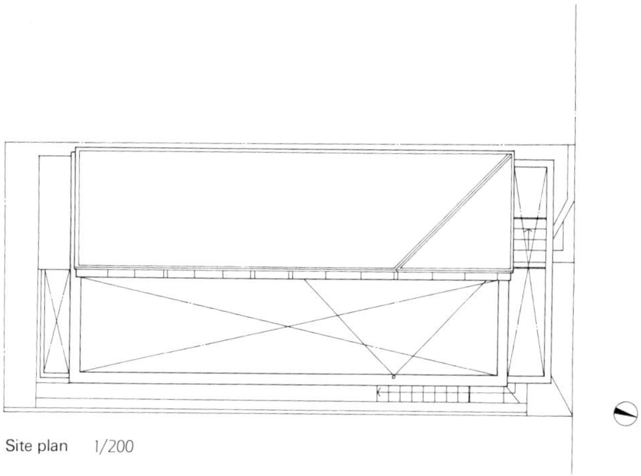

Site plan　1/200

1

2

■The phrase "like the construction that remains" is Mr. Murofushi's motto when it comes to expressing his own architecture. This means that he wishes to produce an ever-lasting scene which resembles the shell of a house after the roof, floors, doors and windows have all been removed.
Many of his constructions have been built with double concrete walls. The outer wall covers the entire building and retains all openings as personal possessions. The inner wall contains all of the openings that it should have, but there is a marked absence of the usual frames, sliding doors and paper screens. In other words, the inner wall exists in direct relation to the external wall and is not affected in any way by the weather or any adverse conditions. It is possible to understand the complete story of Mr. Murofushi architecture by such examples.
The ruined-style wall within is the real wall. It as if Mr. Murofushi had the simultaneous idea to make a production and have fun. The structure wavers between false space and the straight consequences of reality, and makes detours in and out of the house to finally return to the shape of a house. The urge is to create the feeling of an ancient burial chamber and, like an archaeological excavation, produce a construction that is partly concealed, partly open.

■This house is situated in a high-class residential area of Tokyo and stretches well from north to south despite the lack of available width. The concrete rectangular structure has two contradictory aims; one is to adopt a sort of cushion between the road and the house, and the second is to fill in the necessary natural light and ventilation to create delicate features both inside and outside of the house. Additionally, this house was built under the condition that the planning of the rooms was to be strictly controlled by the owner, and furthermore the living room was to be designed for the use of the family and guests. So, this house has been designed with more than one process of thought and consists of separate layers of which some are revealed, and some concealed. For example, the approach where the wall pillars are lined up in a regular manner to the entrance hall, the stairs in the front with the arch ceiling, the living room with its huge Western-style furniture and traditional patterned carpet, the romantic study that resembles a life-sized dolls house, the glass pyramid that creates the connection, the roof top terrace which can be filled with water. All of these features are collected in individual pieces and buried here and there.
It seems as if both the architect and the owner have had great fun with this house with its facilities, furniture and decorations. It is as if the architect himself still lives within the house, concealed as one of the blocks.

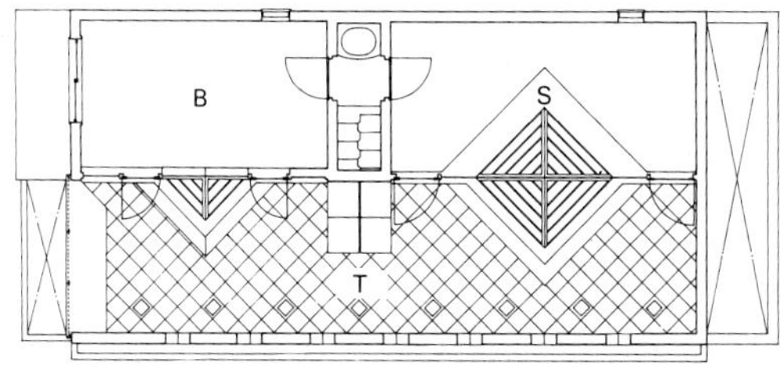

Third floor plan

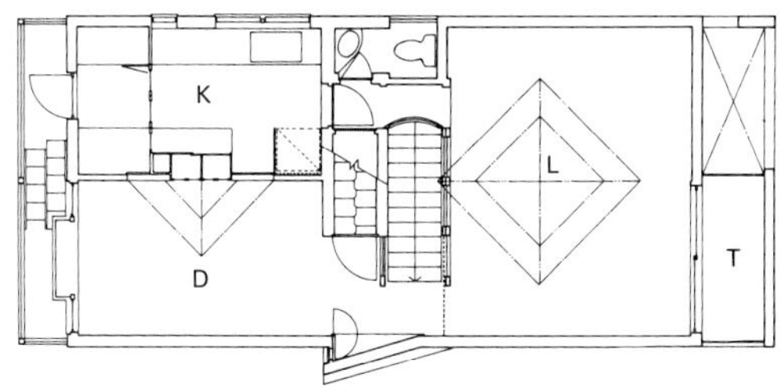

Second floor plan

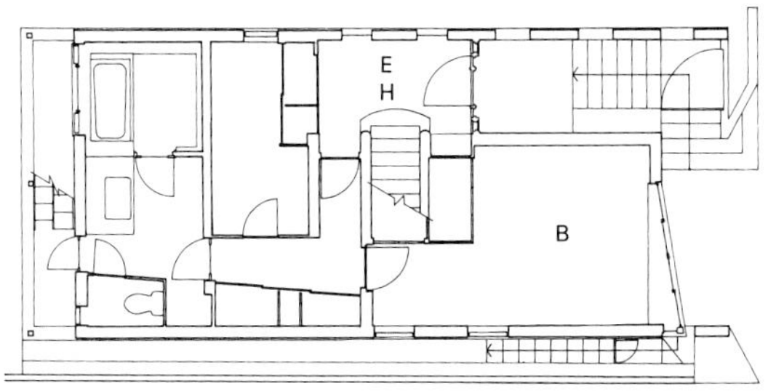

1/200
First floor plan

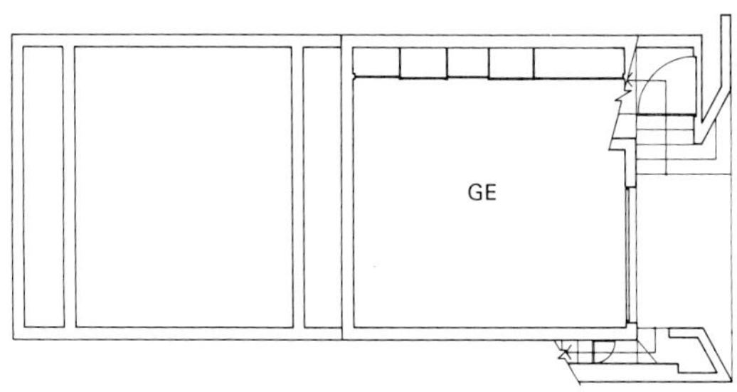

Basement floor plan

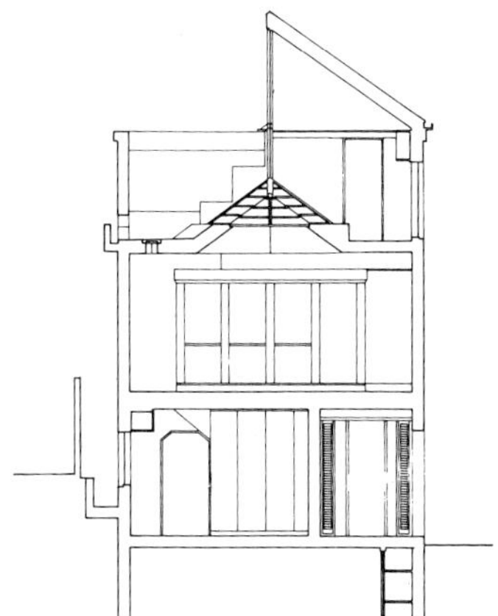
1/200
Section

3

1道路側から見る。階段わきの壁柱は、玄関ホールまで並んでいる。
2階段から玄関ホールを見る。コンクリートの壁柱がここまで並んでいる。
3玄関ホールから2階を見る。左手前に玄関のドア。フォーマルな正面階段とアーチの天井。

1A view from the road. The wall pillars on the side of the stairs are lined up with the entrance hall.
2A view of the entranc hall from the stairs. The concrete wall pillars are lined up with this.
3A view of the second floor from the entrance hall. The front door is to the left of the foreground. The formal front staircase with its arched ceiling.

4

5

6

4階段を上がった所。コンクリートの壁に三角形のアルコーブがある。採光と通風のためでもある。左奥に居間。天井に埋められたガラスブロックも採光のためだが、テラスに水が張られたときは光の状態も変化してくる。
52階居間。巨大な家具やソファやカーペートは最初から決められていた。それに合わせて木と半透明ガラスの階段室の箱をつくり、3階の書斎と視覚的に結ばれたガラスのピラミッドを天井に据えた。
6階段室。右手に居間。居間の伝統的なファーニッシングに合わせて、階段まわりのデザインを決めている。

4At the top of the stairs. The triangle-shaped alcove in the concrete wall. The living room is at the rear to the left. The glass blocks are buried into the ceiling to allow light to penetrate. However, when the roof terrace is filled with water, the light becomes more filtered.
5The living room on the second floor. The huge furniture, the sofa and the carpet were decided upon from the beginning. The wooden and semitransparent stairwell is visually connected to the study on the third floor by the use of the glass pyramids.
6Staircase. To the right, a living room. Design of the stairs and surroundings matches the traditional furnishings in the living room.

7

8

9

10

7 3階書斎から、ガラスのピラミッドを通して2階居間を見下ろす。テラスの表情もピラミッドに連続している。
8 3階テラス。大きなガラス面に割り込んだガラスのピラミッド。奥に、食堂の上部にあるもうひとつのピラミッド。
9 居間天井を見上げる。
10 テラスの中央には噴水があり、全体に水を張ることもできる。

7 View down to the second floor living room through a glass pyramid, from the study on the third floor. The terrace expression is incorporated into the pyramid.
8 The third floor terrace. The large panel of glass at the rear is the glass pyramid. Further back is the glass pyramid over the dining room.
9 Looking up at the living room ceiling.
10 There is a fountain at the center of the terrace that can be filled with water.

土浦の家[House in Tsuchiura]

■

■3世帯3階建ての住宅。若い夫婦と幼児2人が2、3階に、夫婦それぞれの母親が1階に住む。
地盤の悪さに合理的に対処するために、家の中央に集められた鉄筋コンクリートの壁と円柱から左右に梁を伸ばし、そこに木造の部屋を取り付けている。コンクリートの固い外殻のなかの柔らかい部屋という構成を、リバーシブルの外套のように裏返しにしているわけだが、このコンセプトを強調するように、剥き出しになった木造の壁を薄いガルバリウム鋼板で包み、「裏地」として見せている。外観は、それぞれの部屋の要求が、不規則な形態や開口部や装置をつくり出し、あえて雑然とした様相を呈しているが、コンクリートの壁と柱と梁は、ヤジロベエのようにシンメトリカルにバランスを保っている姿を、妻側の立面に現している。
だから、人々は住宅の内核で構造体と出会い、一緒に暮らしている。1階では2枚の壁は露地のような吹き放ちの通路をつくり出している。この壁を外側の囲いとして、通路の両側に2軒の住まいがある。この壁は2、3階においても立ち並ぶ壁柱として、開放的なそれぞれの部屋の間の緩衝帯的な通路になっている。さらに、どの部屋のなかにも独立した円柱が顔を出す。強い形で日常のなかに現れる構造体が、生活のよりどころにもなっている。

■A three-story house occupied by three generations. The young couple with their two children occupy the second and third floors, and the first floor is occupied by their mothers.
In order to deal with bad foundation conditions, a ferroconcrete-based wooden room has been constructed in the center of the house, and a beam has been stretched to both sides of the columns. The constructional concept of this house is that of softness within a hard concrete shell but which has the impression that it can be turned inside out like a reversible coat. To emphasize this concept, the bare wooden wall has been thinly covered with a coat of galvanoplastly to achieve the effect of a lining. The external appearance of the house is a little disordered owing to the irregular shapes of the windows and doors which were created to meet the demands of each individual room. However, the concrete walls and pillars tend to maintain the balance. So, the owner's family compromise with the internal construction and live happily with it. The effect of the two sheets of wall creates an aisle like an unoccupied lot. There are two houses to the west of this aisle, so it also acts as an external fence for the first floor. This aisle also acts as the neutral zone between the open rooms on the second and third floors. The wall pillars, which stand in line, can be seen in each room and change the complete structure into a saucer for daily life.

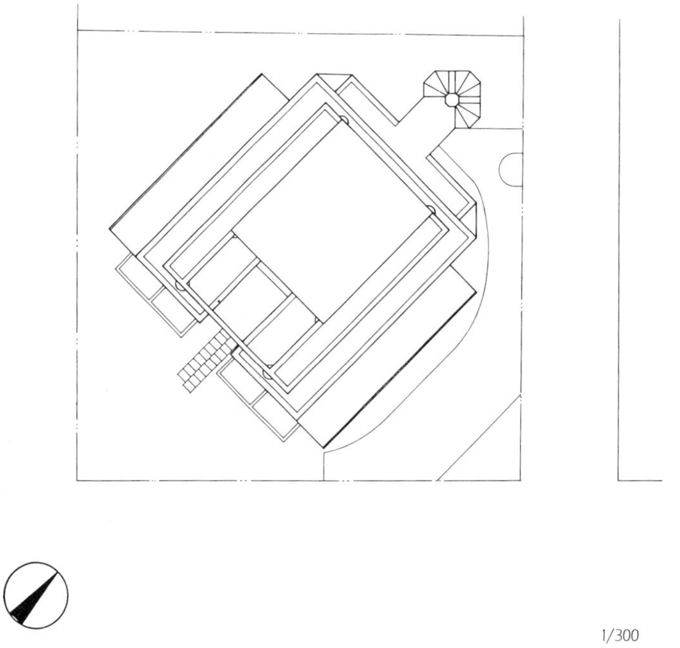

1/300

Site plan

■■南側外観。交差点の一角地に建っているが、建物は敷地と対角に、交差点に対して平行に向けられている。

■■An external view of the southern side. The house is situated in a corner of the junction, but lies parallel to it.

2

3

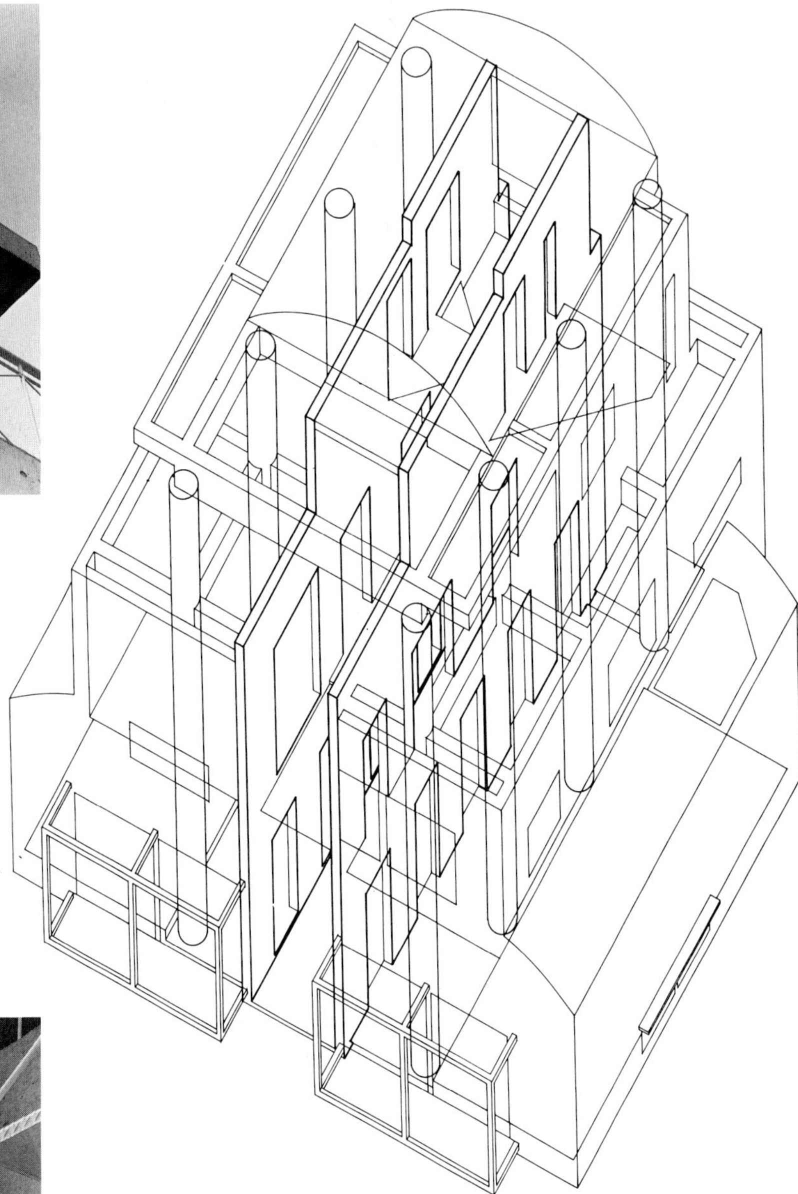

4

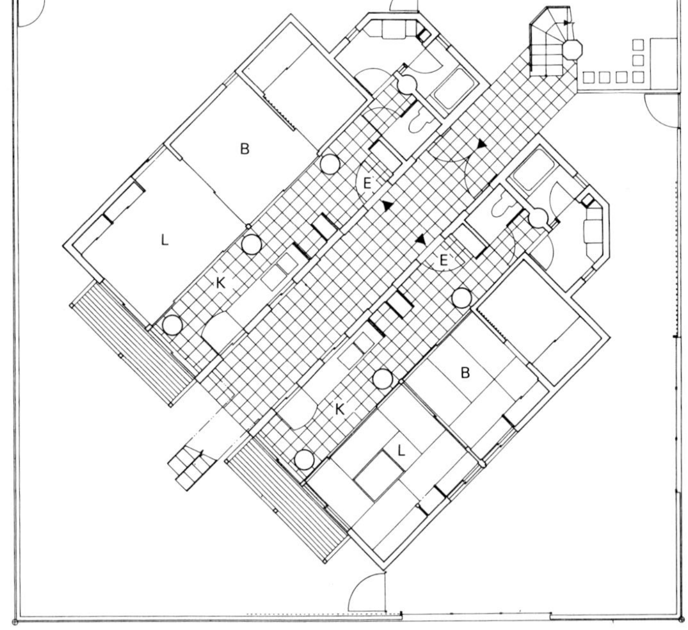

First floor plan 1/200

34 南側ファサードのディテール。コンクリートの構造体はシンメトリー、それに、ガリバリウム鋼板、アルミサッシュ、木のパーゴラ、鉄骨階段などが、自由に絡まる。
5 北側外観。若夫婦の住居入口が、階段を上がりきった2階中央に見える。南側にも直接庭に下りる鉄骨階段がある。

34 Details of the south facade. The structure is symmetrical concrete, but a galvanized iron sheet, an aluminium sash and wooden pergola can be seen clinging to it with ease.
5 An external view of the northern side. When climbing up the stairs to the second floor, an entrance to the house used by the young married couple is visible in the center. On the south, there is another steel-frame stair going directly down to the garden.

5

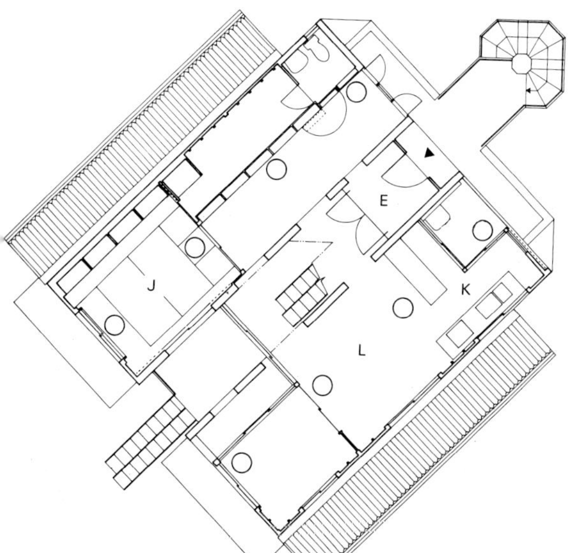

Second floor plan

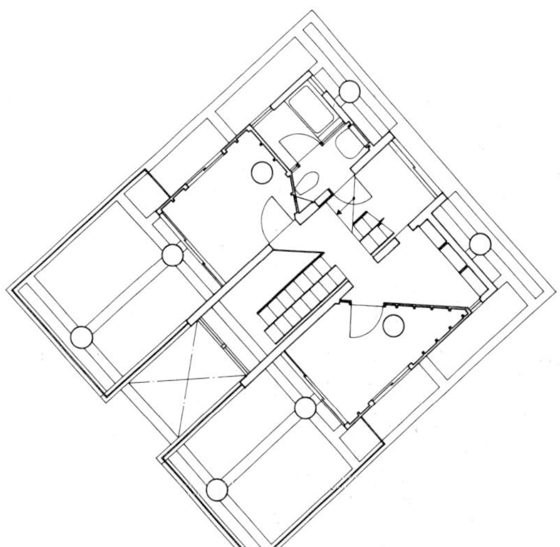

Third floor plan

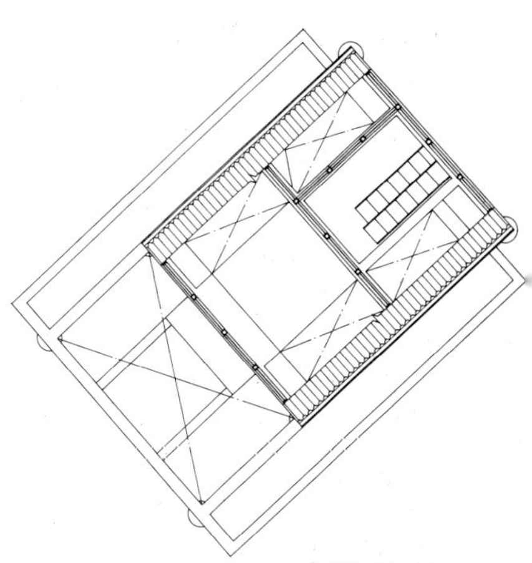

Fourth floor plan

6

7

6 2階。玄関わきの細長い部屋(書棚が並ぶ夫の部屋)から、階段室のある通路と奥の和室を見る。コンクリートの円柱と畳、障子の取合わせ。
7 2階。食堂から通路を通して夫の部屋を見る。8本の円柱があちこちに独立して現れる。
8 3階の階段踊り場から、中央通路の南側を見下ろす。上部に屋根裏部屋。

6 The second floor. A view of the passage with the stairwell and Japanese room beyond it. The long room (crammed with bookcases) is on this side of the entrance. The contrast between the combination of concrete columns, tatami and paper sliding doors.

7 The second floor. A view of the husband's room through the passage from the dining room. The eight columns emerge from everywhere as totally independent entities.
8 Looking down the south side of the central passage from the landing on the third floor. The attic is above it.

8

9

10

11

9 2階から階段を通して3階、屋階を見上げる。家の中央にコンクリートの2枚の壁が立ち、そこから梁が外へ向かって腕をのばしている。
10 2階玄関ホール。
11 3階踊り場。
12 2階和室。その外のグレーチングの床が南側の外階段とつながる。
13 2階和室より、グレーチングの床を通して反対側の作業室を見る。
14 下から南北に吹き放ちになっている中央通路。北側の玄関ドアもスチールメッシュを張っただけ。通路に面して両母親の住まい。通路は染色家の夫人の作業場としても使われる。

9 Looking up to the third floor and roof from the second floor stairwell. The two sheets of concrete wall stand in the center of the house, and the beams stretch outwards from there.
10 Entrance hall on the second floor.
11 The third floor landing.
12 The Japanese room on the second floor. The grating outside of this room continues to the outside stairs on the southern side.
13 The work room is viewed through a grating from the Japanese-style room on the second floor.
14 The central aisle left open to the south and north. The entrance door in the north is only covered with steel-mesh. The mothers' living quarters stand facing the road. The passage also acts as a workyard for the wife, whose occupation is a dyer.

12

13

14

内藤 廣 [NAITO Hiroshi]

共生住居 [The Commensally House]

1

2

■内藤廣は、自分の手がける建築をできる限り単一の要素の反復に還元しようとする衝動にかられている。単純な形式を追いつめながら、あまりにも(彼自身がいうように)偏執的であり執拗であるために、どこか過剰な形式として溢れ出てしまう。例えば、栃木県那須に建つ広大な住居は、大小24枚のフラット・ルーフの重なりからなり、これらのコンクリート・スラブは、屋根が窓台に、ついには床になっていく。典型的な例といえる。
彼は、予測しようのない生活の多様な姿に魅せられているので、その容器である建築を、家の形になった時点で手を休める理由をもたないのである。那須の家では、積層するスラブが自然と融合する地点が夢見られている。東京・品川に建つ住居は、整然と並ぶコンクリートの壁柱が家の輪郭を平面的にはまとめているが、その他の部分は、多様な素材、形態、色彩によってアドリブ的に埋め尽されている。こうした部分部分の意匠は、内藤だけでなく、スタッフ、石工、大工、左官工、板金工、施工などの参加によってもいる。またある住宅は、住み手がさまざまな植物を持ち込むことを予想してつくられている。生活との出会いが、建築の変容へと爆発することへの期待。

■コンクリートの壁柱が上下階に並んで立ち、長手中央部分が吹き抜けている。細長い空間がくびれたような形で連結している部屋は、どう使いわけても構わない。現在は吹き抜け部分の中ほどを仕切って、中央部を2世帯それぞれの居間にしているが、2階では9つの部屋を連鎖状に、半ば開いたままでつなげている。間仕切を通して隣りの気配も感じられる、ゆるやかにわけられた、4世代家族の共同生活の住居である。
このように半ば開き半ば閉じた連続的空間をつくり出すためと、ローコストを目指して、構造は鉄筋コンクリートの薄肉ラーメンにしている。型枠工事の費用も惜んで、梁を出さないようにし、建具用のスチール・レーンや断熱材のドリゾール板をフラットな型枠の上に敷き並べてコンクリート打ちを行っている。内部間仕切は厚さ18mmのラワン材を積み上げ、残りは可動の壁としての木造建具。いつでもバラせる状態になっている。
すべてがぎりぎりのところまで詰められ決定されている。無駄な部分を一切削り落としているが、その建築の印象はホットである。その最終目的は、形式にとらわれない生活の全面的な展開の場になることだから。

3

1 庭側から見る。2階の雨戸を完全に閉めると、床スラブは隠れてしまう。
2 2階。オープンな部屋が長平方向に連続し、コンクリート壁で分節される。
3 4 1階、居間2。もう1つの居間とを仕切る棚風の間仕切は18mm厚のラワン。

1 A view from the garden. The shutters on the second floor conceal the floor-slab when closed.
2 The second floor. The open room continues along its length and is separated by a concrete wall.
3 4 The first floor. The second living room. The shelf-style lauan dividers separate this room from the other living room.

4

5

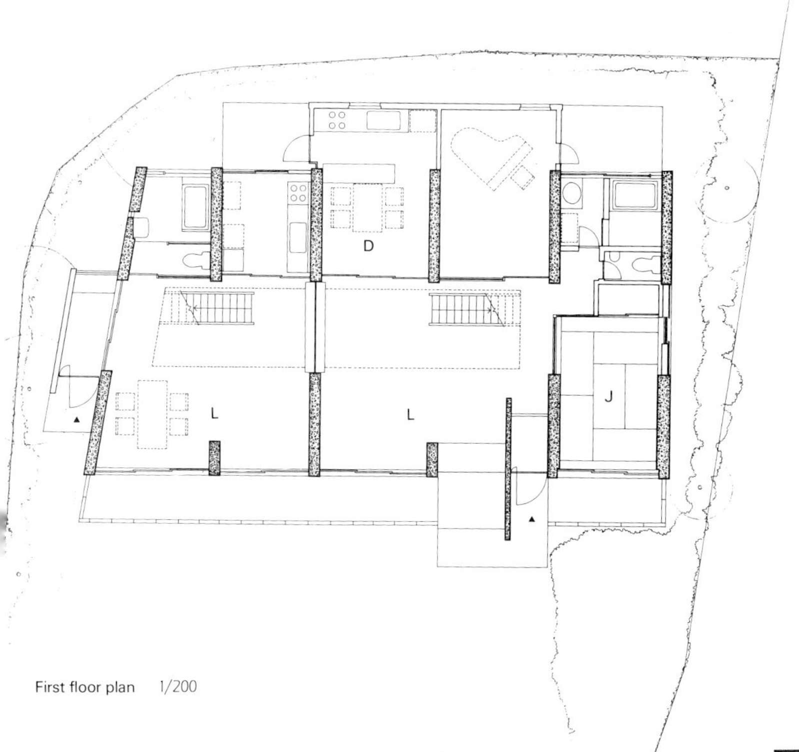

First floor plan 1/200

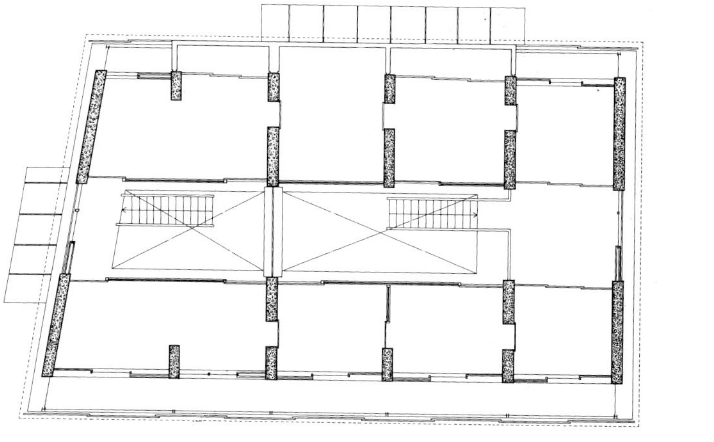

Second floor plan

■Mr. Hiroshi Naito feels the urge to express his architecture with the repetition of as many single elements as possible. According to himself, he is obstinate and persistent and this somehow leaks out in his excessive style. The enormous building in Nasu, Tochigi prefecture, is a perfect example of this style. It is constructed by twenty-four layers of concrete slabs which act as roofing, window holders, and finally reach the ground.
He is attracted with so many aspects of life that he feels no compulsion to withdraw his hand once the actual shape of the house has been achieved. The house is Nasu gives off a tense harmony with nature in the way the slabs have been piled up. His designs in Shinagawa, Tokyo have been flatly arranged of concrete wall pillars which add a feeling of order to the structures, but the remainder of the structure has been buried amongst the use of various materials, formations and colors in an ad-lib fashion. Such abstract design is also owing to the fact that Mr. Naito allows the staff, the masons, the carpenters, plasterers, metal workers and construction workers a say in the production. It has been created with the anticipation that the new tenants will bring with them an abundance of plants. It is the basic dream that a life-style can alter the appearance of the structure.

■The concrete wall pillars stand side-by-side on the upper and lower floors, and the central area stretches up to the ceiling. The room connected to this compressed area can be used for any purpose. The central area of the first floor has been divided into two to provide a living room for two separate households, but the nine rooms on the second floor have been left half open as they are all connected to one another. It is possible to hear the conversations in the neighbouring rooms through the dividers. This is a gently divided house for the use of a four-generation family.
Rahmen ferroconcrete was chosen to create a semi-open feeling to the house, but also to keep the costs low. A further saving has been made by the use of the steel frame and heat-resistant materials being placed on top of the flat mould into which the concrete is poured. The 18mm planks of lauan have been piled up to separate the inside, and wooden fittings have been used to act as a removable wall that can freely be taken out.
The minimum of expenses have been used in the construction of this house. Although all unnecessary elements have been eliminated, the house still manages to give off a warm feeling. The final aim is to create a full scale house for a life-style that is free from formalities.

6

5居間2。左手前にダイニング・ルーム。I型鋼の中に照明器具を入れている。
6庭側(南西側)ファサード詳細。ステンレス・ワイヤーの手摺が、コンクリート・スラブの水平性を際立たせている。

5The second living room. The dining room is on the left foreground. The light fittings in the I-beam.
6The detail on the garden facade (south-western side). The stainless wire railings emphasize the vertically of the slab.

7

8

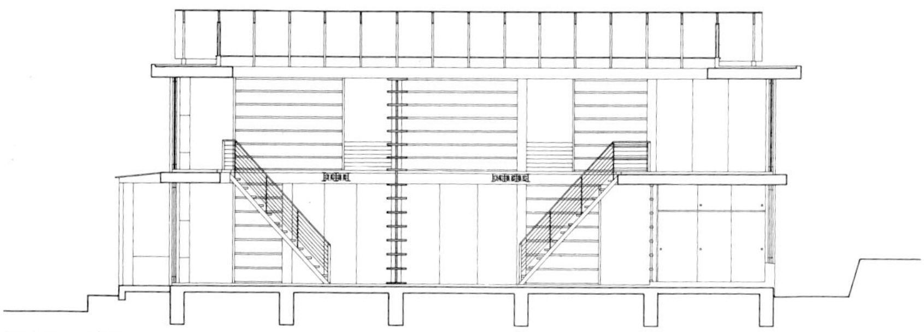

Section　1/200

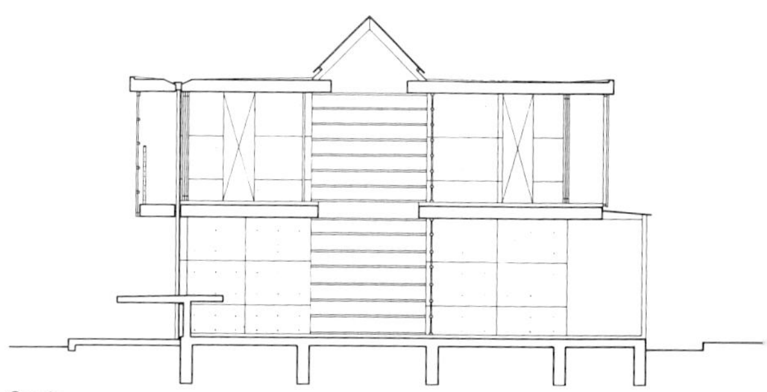

Section

7居間2。コンクリート壁柱の奥に、内外を貫通するもう1枚のコンクリートの壁が、玄関部分を形成する。
8 2階。室内より縁側を見る。天井のコンクリート・スラブに、建具のレールとドリゾール板が打ち込まれている。
9 2階。庭に面した縁側。雨戸は天井のレールから吊られている。

7 The second living room. Another concrete sheets that runs through to the back of the concrete wall pillar, make the entrance.
8 The second floor. A view of the verandah from indoors. The fitted rail has been fixed to the verandah roof.
9 The second floor. The verandah facing the garden. The shutter is suspended from the lace on the ceiling.

9

10

10居間の吹き抜けに面した個室。間仕切壁はすべて仮設的な木造。
11棧木をはずせば、このラワン板もすべてバラバラにはずれる。
12庭から見る。

10The private room facing the living room with its high ceiling.
11The planks of lauan will become removable if the shelves are taken out.
12A view from the garden.

11

12

六角鬼丈 [ROKKAKU Kijo]

塚田邸パートII [Tsukada House Part II]

■六角鬼丈は、1970年代のはじめに一連の自邸の計画を発表している。そのひとつは放心気、遊心気、求心気と名づけられた、球形、環形、円錐形の原空間が建物のなかにくり抜かれたものであり、また他には、「伝家の宝塔」と命名された、家族の記録を収納する巨大な家具が、家の中央部に据えられたものだった。いずれも半ば幻想建築といえる計画であるが、機能追求によってすべてが決定される現代住宅の計画手法に対して、精神的なよりどころとしての空間や記憶の問題に対応する装置を建築化した例としては、異例といっていいほど早く、また完成度もきわめて高かった。
しかも、六角の手掛けてきた住宅のなかに、この原空間や装置が、何らかの形で現実化されてきている。家のなかにすっぽりと入ったドーム状の部屋、パースペクティブな階段室、家の中央部を真円でくり抜いた中庭、等々。それらはつねに奇想のような見え方で実現されているが、建築家の精神が一貫して、明快な形で建築化されているまれなケースである。
これらの原空間は、その形状からして可塑的な素材を必要としているように見えるが、例えばシュタイナーの建築のように直接的な表象化を六角は避けている。静的な表現主義的な建築ではない。むしろ普通の家のなかに割って入る。家は溶解過程を経て、精神宇宙のモデルへと変貌しはじめる。

■1980年に、六角が設計、完成させた住宅の増築である。
その母屋は、一見ごく一般的な木造住宅のなかに、3本の赤松の根付丸太が貫入したものだった。とくに中央部に据えられた丸太は、居間の炉端に根を下ろし、家のなかを斜めに抜け、2階の子供部屋のところで梢の部分が広がっている。このような丸太は、建築の装飾的要素ではもはやなく、建築そのもの以上に建築軀体に近いものだろう。
この増築棟(アネックス)は、木に対して溶岩をもち込んでいる。たった一間(ひとま)だけの建物であるが、海に見立てた黒い床、天空としてのコンクリートの曲面天井と壁、その間に盛り上がり、流れ出している溶岩の島を見れば、そこは部屋であると同時に庭園でもあるだろう。屋根の上にはもうひとつの現実的な庭――温室がのっている。
このコンクリートの箱は、住宅というよりは天地を閉じ込めた現代の厨子である。建築から家具あるいは什器に変わりつつある。曲面天井に沿って切り開かれた眉形のハイサイド・ライトは、そのような衝動を十分に語っている。
母屋の松丸太も、この離れの溶岩も、あくまで重く強い。しかし物質そのものの重さではなく、精神の重さ――つまり建築の内部においては、生きている形のままの木や岩は、加工された素材よりも、むしろコンセプチュアルな存在なのだ。

■The plan to design a series of houses was announced in the early 70's. One of the basic themes of the constructions were to be balls, circles and cylinders which in turn were named "absent-mindedness", "playfullness" and "desire". These shapes were to be incorporated within each structure. The second theme was the use of huge furniture that was to be handed down through the family as living history books and called "the hereditary treasure tower". These were to be placed in the center of the house. Both of these plans seem pretty fantastic, but in comparison to modern architecture where the room needs to fit the function of the people who occupy it, in this case the end product that was achieved was the expression of the equipment used to denote a spiritual reliability which correctly utilizes the available space and still manages conjure up memories of the past. This is a simple way of explaining the phenomenon, but there was obviously a high chance of success contained within the idea.
Furthermore, the basic use of space and equipment that is contained in this idea has been successfully used before by Mr. Rokkaku in the dome-shaped room that was inserted in its entirety into a house, the perspective staircase, the garden that was placed into the core of the house, etc. These points that were actualized took on a sort of fantasy aura, but it is very rare for the architecture to achieve the expression of the consistent spirit of its architect in such clear shapes.
The basic area used seems to need earthy materials and shapes, but Mr. Rokkaku makes a point of avoiding the direct shapes that are revealed in the architecture of Rudolf Steiner. However, his work is not static. It tends to lend itself easily to ordinary houses. The houses went through the process of being dissolved and were then transfigured into a caricature of space.

■This building is an extension of the main building that Mr. Rokkaku designed and completed in 1980.
Three Japanese red pines complete with roots were passed through an ordinary wooden housing area to make up the main building. In particular, the log placed in the center places its roots beside the fire-place and spreads is branches diagonally through the house until it reaches the children's room on the third floor. In this way, the log changed from being a mere decoration and actually become a part of the architectural structure of the building.
Lava has been chosen for this annex as opposed to trees for the main building. Despite the fact that it is only one room, the lave flow of the black floor and the wavy feel to the concrete walls and ceiling makes it appear like an island of lava and could be used as a room and a garden. There is another garden in the shape of a greenhouse on the roof.
This concrete box is a modern miniature shrine which, it seems, is not so much a room but a place with the universe closed out. The interior architecture is directed by the furniture or instruments available. This impulse is completely summed up by the eyebrow shaped light

1

fitting which is located in the ceiling.
The pine tree logs used in the main building and the lava used in the annex are very heavy-duty and strong. However, it is not the weight of the materials that is apparent, but the mental weight that the live trees and living rock add to the house, something that processed materials cannot provide.

2

3

1室内東面。天空に見たてた曲面天井。正面収納の扉に描かれた宗達の絵を模した風神雷神は漆、床の間の水源を主題にした掛け軸は刺子。どちらも施主家族の手になる。
2南側外観。屋上は温室。
3西側外観。内側の床から溶岩が流れ出している。

1The eastern side of the room. The wavy ceiling which is like the sky. A copy of "Fujin-Raijin" by Sotatsu has been painted on the door of the closet and finished in lacquer. The hanging picture in the alcove is embroidered. Both are the work of the family living in the house.
2An external view of the southern side. The roof is a greenhouse.
3An external view of the western side. The lava seems to flow from the floor within.

4

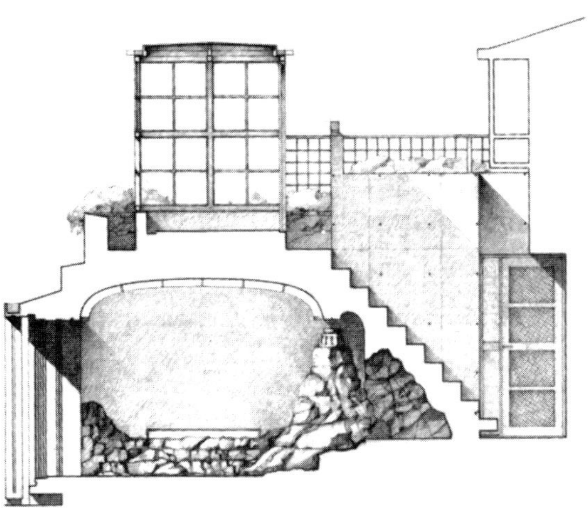

Section 1/100

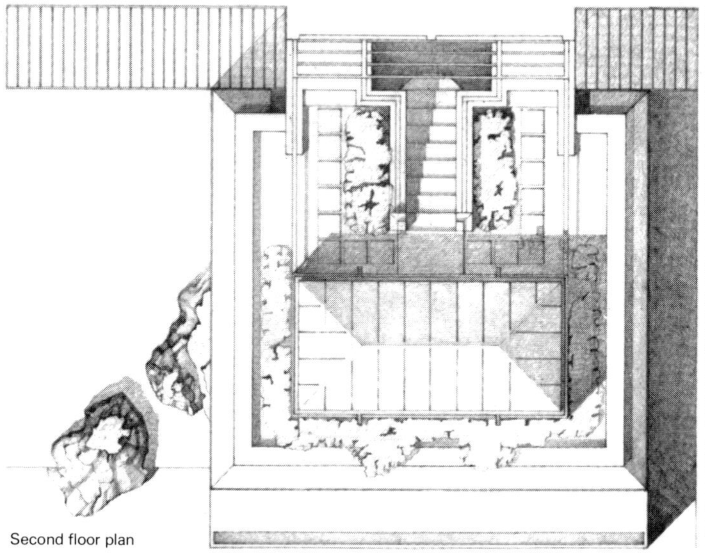

Second floor plan

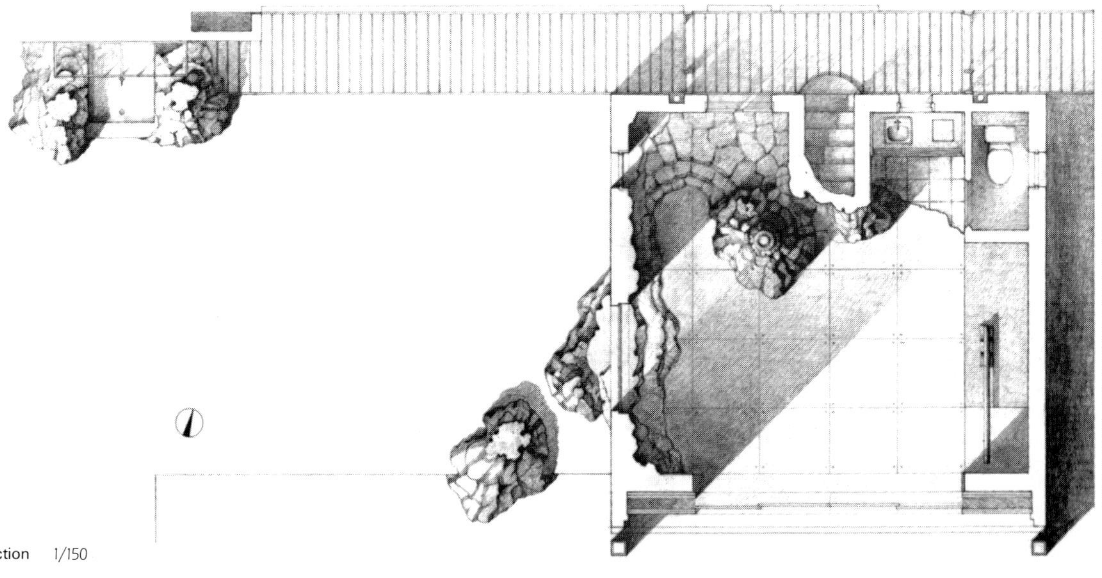

First floor plan Section 1/150

5

6

4室内北面。右端に便所、その左にキチネット。溶岩の奥に入口。床はブナ合板敷込みのうえ、黒のウレタン塗装。家族で焼いた七宝焼が埋め込まれている。
5室内西面。日光のさし込む天井のスリットと溶岩が流れ出す床のスリット。溶岩噴火口頂部の彫刻も施主の手加工。
6西側外観。

4The northern side of the room. The lavatory is on the right. The kitchinette is on the left. The entrance is on the other side of the lava flow. The floor is covered with a sheet of beech and painted in black urethane. The cloisonneware, which has been made by the family, have been placed here and there.
5The western side of the room. The slit on the ceiling allows natural light to filter in, and the slit on the floor is for the lava flow. The statue which stands on top of the mouth of the lava flow was also made by the owner of the house.
6An external view of the western side.

青葉台の家[House in Aobadai]

1

■全体の輪郭は、箱形の上にシリンダーをのせ、そのまま2つに割った形。しかもそのシルエットをなぞる線がフレームとなって際立たされている。といって、この異形のフレームは構造体を意味しているのではない。壁構造である。フレームのなかに充填された壁なのではなく、構造壁を縁取る飾りにすぎないフレーム。だから道路側ファサードで印象づけられる孤立したフレームは、私たちに構造体の輪郭を見せているのではなく、生活領域の輪郭を示しているのである。
いかにも六角らしい転換である。物質の真実性、とくに打放しコンクリートの軀体は、その真実性を端的に見せやすい素材だが、六角はコンクリートを用いながら、その属性を切り離している。素材の真実性は、現実の家の形をつくり出すが、その家の形は、住宅地のなかで個の領域を決定的に切り取り、囲ってしまう。そうした住宅の存在形式に対して、六角は、半ば透けたポーラスな家をここで示そうとしている。
フレームごしに垣間見える庭、門のドアによって柔らかく仕切られたアプローチ。外階段によって心理的に外部に誘い出されている2階の居間、そしてユーモラスな形で家らしさを見せている全体の輪郭、等々。この住宅は四囲の環境のなかに溢れ出している。

■This house is in the shape of a cylinder that has been placed on top of the rectangular foundation and then cut in half. The line that draws the silhouette has been made to stand out as a frame. However, the most unusual part is not the shape of the building structure.
But the structure of the wall. It is not more than the frame which decorates the boarder of the party wall. In this sense, the independent frame on the facade is not showing the outline of construction, but it is indicating the territorial rights that comes with living here. This is a typical turnover of Mr. Rokkaku. The reality of the materials used, particularly the ferroconcrete, tend to reveal its true features. However, despite the use of the concrete,

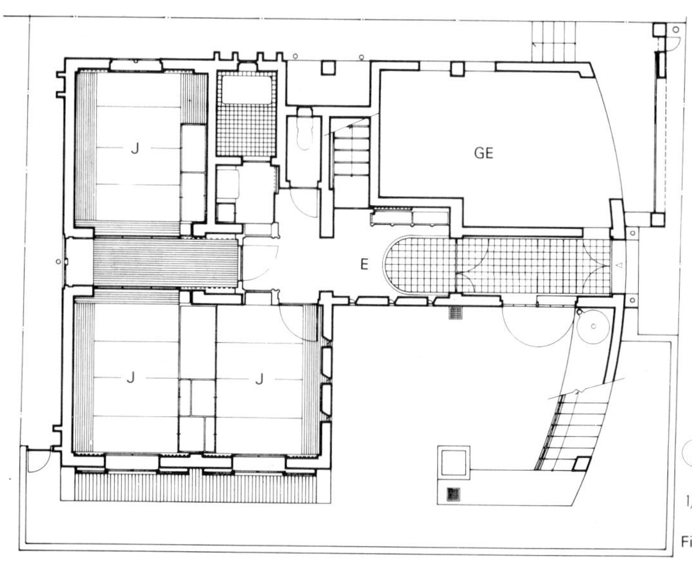

1/150

First floor plan

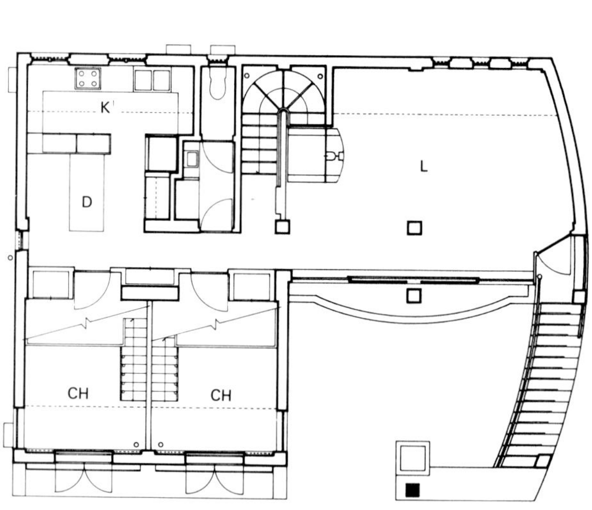

Second floor plan

the espression he has achieved is totally independent to its attributes. The reality of the materials produce the actual shape of the house. And this shapes cuts out its own individual territory in the housing area and confirms it. In opposition of the substantial housing thereabout, Mr. Rokkaku tries to express a feeling of semi-transparency and porousness into the house.
For example, the garden that can be viewed through the frame, the approach that has been gently partitioned by the gate, the second floor living room that has been designed to psychologically lure the occupant outside by the effect of the outside stairs, and, in fact, the whole humourous aspect of the house in profile has managed to leak out of the rectangular confines of the house.

2

3

4

1道路(東)側のファサードと、そのフレームと外階段によって囲われる南側の庭に面した外観。
2東側ファサードの見上げ。輪郭はシンメトリー。
3門の扉。
4東側のファサードの一要素になっている外階段から、2階の居間に直接アクセスできる。階段はフレームに沿って弓なりに曲っている。

1The facade on the eastern side and an external view of the southern side of the garden which has been confined by its frame and outside stairs.
2Looking up at the eastern side facade and its symmetrical outline.
3The gate door.
4Access to the second floor living room is available from the external stairs that are one of the elements that go to create the eastern facade. The stairs curve along the frame like an arrow.

5

6

7

8

9

5南側外観。2階バルコニーを見上げる。道路側ファサードの湾曲した曲線が、もう一度繰り返されている。
6南の庭を囲んでコの字形の建物。
72階居間から張り出ているバルコニーのプラント・ボックス。弓形のハイサイド・ライト。
8南側から子供室を見上げる。ここにも湾曲した庇が現れる。窓は両開き。
9玄関ホールから門の方向を見る。

5An external view of the southern side. Looking up at the balcony on the second floor. Here the curved facade of the road side has been repeated.
6The U-shaped building that surrounds the garden to the south.
7The plant box of the balcony projects out from the living room on the second floor. The arc-shaped high side lights.
8A view of the children's room from the south side. The curved cover appears here again. The double swing windows.
9A view of the gate from the entrance hall.

10

11

12

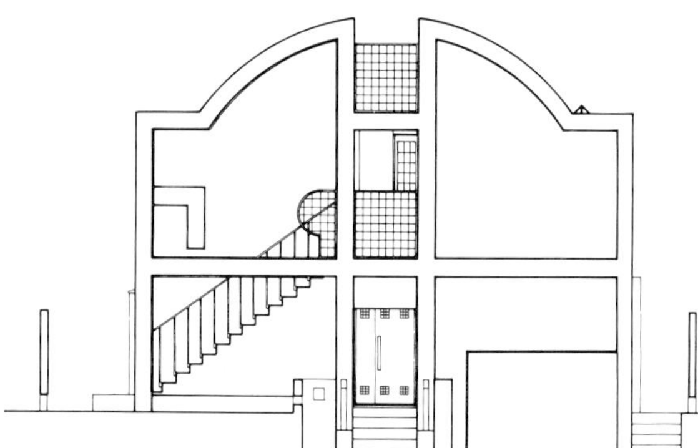

East elevation 1/200

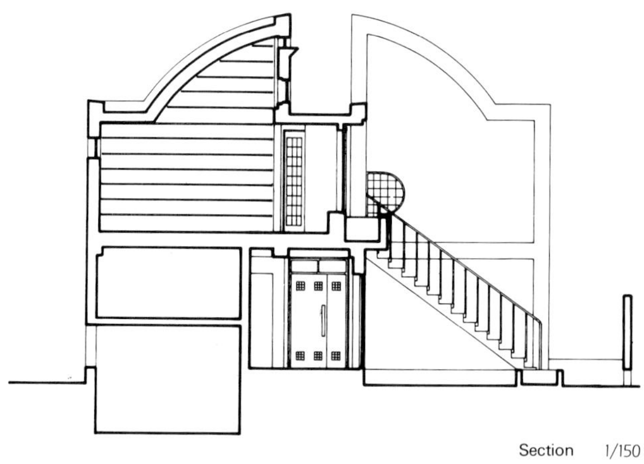

Section 1/150

13

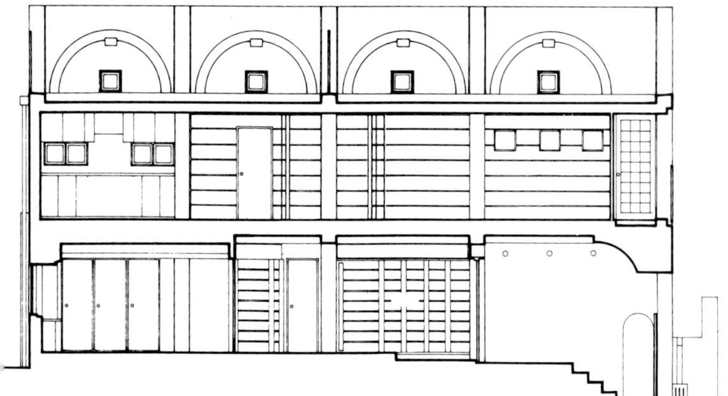

Section 1/150

10居間から食堂方向を見る。左手に子供室。突き当りに正方形の窓が縦に並ぶ。この正方形は、他の開口部における格子の基本パターンでもある。
11階段室。右手に居間。
12和室の開口部。
13居間。この住宅の模型のような箱家具は、位牌や家族のビデオテープなど、家の記録の収納装置である。

10A view of the dining room from the living room. The children's room is on the right. The rectangular windows are lined-up at the rear. This square shape is also the basic pattern of the grill that represents the other openings.
11The stairwell. The living room is on the right-hand side.
12The door to the Japanese room.
13The living room. The box-shaped piece of furniture that resembles the shape of the house holds the family records such as ancient documents or recent family videos.

14

15

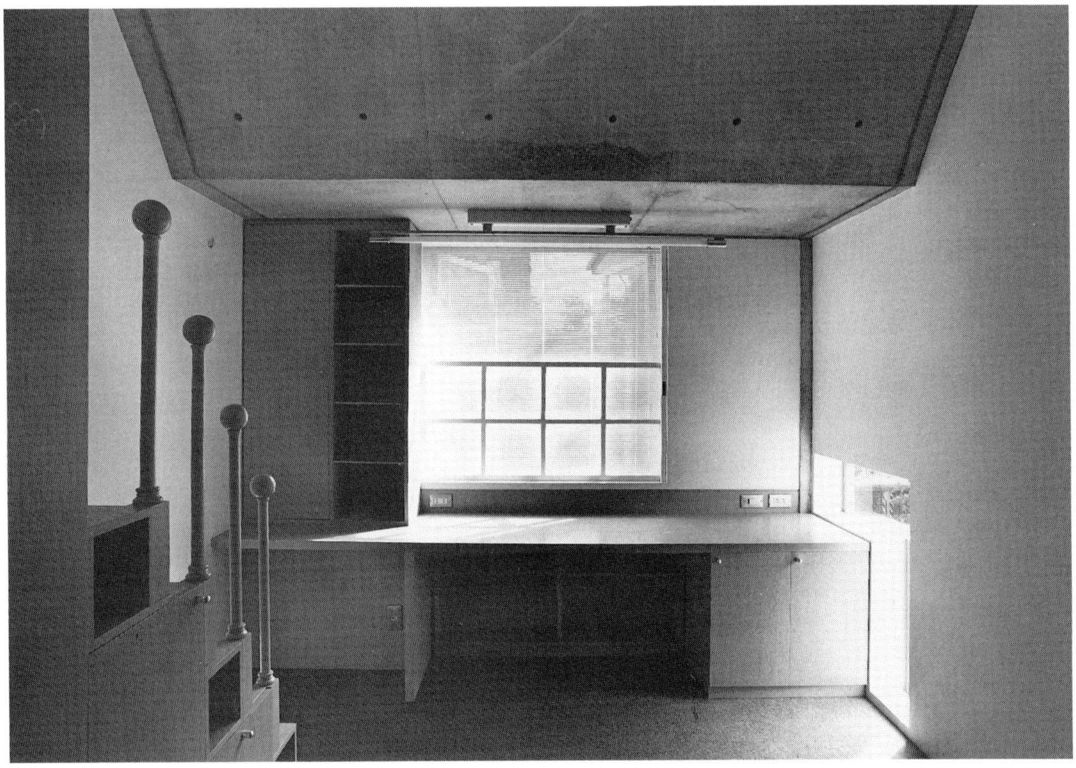

16

17

18

14居間のハイサイド・ライトと半ヴォールトの天井を見上げる。
15居間。外階段に出るドアが見える。
16子供室の南面を見る。右手前にロフトへ上る階段。
17子供室上部のロフト(寝床)。
18居間・食堂のハイサイド・ライトと子供室のロフトが、外で顔を合わせている。

14Looking up at the high side lights and the semi-vault on the ceiling of the living room.
15The living room. The door leading to the stairwell can be seen.
16A view of the southern side of the children's room. The stairs to the left at the rear lead to the loft.
17The loft (bed) above the children's room.
18The high side lights for the living room, dining room and children's loft face each other on the outside.

鈴木 恂[SUZUKI Makoto]

ATL 8110

■鈴木恂は、大学を出たあと、世界を旅して帰ってくると、そのまま自分の設計事務所をつくって活動を開始した。1960年代後半に最初期の住宅が集中している。いずれも、打放しコンクリートの箱に、ガラスの開口部が直接取り付けられたような、強烈な原始性が感じられる。
ある意味では考えられる限り単純化されたもののもつ原始性である。同時に、それは彼がメキシコや南米の民家、スイスの集落や北欧の山間部の家々、イタリアやモロッコの道、エーゲ海の島々を単独で歩きまわり、実測してきた体験を含んでいる。さらに、雲や水を長年にわたって観察してきた目が、そこに重ねられている。またさらに、ル・コルビュジエやヨルン・ウッツオン、中南米の現代建築家たちの仕事を詳細に研究した成果もそこに取り入れられている。そのような原始性であり、単純さである。
ここから透視されてくるのは、現代建築が、住む形を原型に立ち帰って再構成した、20世紀においてもっとも充実した時期の、理想とされた住居の姿である。そこには住む者の住み方が、席を空けて待たれている。鈴木のコンクリートの箱は、住むことを待つ姿である。

■小さな路地に面して建つ、建築家のアトリエと書斎、ピアニストである夫人のスタジオとしてつくられた小さなコンクリートの建物。鈴木自身のアトリエであり、近くにある彼の住まいの「創造的部分だけのための」アネックスとして考えられている。
不整形の敷地と斜線制限とがそのままプランと外観に表れている。制約の多い条件のなかで、地下階から地上2階をまとめる柱梁を基本とした、いわばプリミティブ・ヒュッテであるが、2階の上部に高い屋根を伸ばし、ある部分では壁を二重にとって光や風の抜けるアルコーブをつくり、どこにいても家全体の垂直的な伸びと、屋外の家々や道や自然との結びつきを感じさせる、繊細な工夫を凝らしている。
内外ともに打放しコンクリートのまま、大きな開口部のガラスはすべて嵌め殺しであるが、いかにも手づくり的な柔らかい表情が、小さな路地にふさわしい親しみのあるものになっている。
いいかえれば、建築ではなく、装置そのものであり、鈴木の装置への志向が幸福な私性を得て領域化されている。建築の社会性がつねにものの力、例えばコンクリートを明示しようとしてきたのに比べて、鈴木は私性によってコンクリートを必要最小限のシェルター、いわば世間には見えない壁であることを望んでいる。

■Mr. Makoto Suzuki, after graduating from university, travelled around the world, and immediately after returning, established his own architectural design firm. Most of his early houses were designed in the late 60's. One feels a dazzling primitiveness in his buildings, each of which is a ferroconcrete box with glass openings to the outside.
This is the primitiveness belonging to the most extreme simplification. It reflects his experiences of lone wandering through private houses in Mexico and South America, villages in Switzerland, houses in the mountains of northern Europe, roads in Italy, Morocco, and Aegean islands. It is also influenced by his observant eyes for clouds and water, and incorporates his detailed studies of Le Corbusier, Jørn Uezon and modern Latin American architects. These are the ideal houses which came out of the fruitful years when modern architecture returned to the original form of the dwelling. Mr. Suzuki's ferroconcrete box waits for a dweller to arrive bringing his own way of dwelling.

■A small concrete building facing a narrow lane, constructed as an atelier and study for the architect and as a studio for his wife, a pianist. Mr. Makoto Suzuki regards this atelier as an annex to his house nearby, to be used "Solely for creative purposes".
The plan and appearance indicate an irregular plot and oblique line limitations. Built under these restrictive conditions, it may be called a primitive hut with a pillar-beam structure which units the basement and two above-ground floors. It was delicately planned to create a sense of vertical stretch anywhere in the house, with a high roof on top of the second floor, and with an alcove for light and wind formed by double walls, which ties up the connection between nature, the road and the house.
It has ferroconcrete surfaces inside and out, and large glass openings fixed in its sides, but the feeling of careful and individual workmanship gives it an intimate atmosphere suited to a narrow lane.
This is not a building, but a setting. And Mr. Suzuki has given it a happy feeling of privacy. The fashions of Architecture have supported open displays of bare concrete, but Mr. Suzuki uses it only out of necessity. It is a minimum shelter for privacy, an invisible wall, we might say.

1

■コンクリートの壁面をすっぽりと蔽ったツタが、大きな開口部を別の表情で浮かびあがらせる。
■南面からの全景。細い路地に、いかにもくつろいだ感じで開かれた家。右手前は同じ建築家による、もう一軒の住宅。
■プリミティブ・ヒュッテに近い素直な形をまとめたこの家に対して、手前の住宅はやや演劇的なポーズで決めている。

■Ivy covering the concrete walls, frames the prominent openings.
■A complete view from the south side. An open house with a relaxed atmosphere. Another house designed by the same architect, right foreground.
■In contrast to the primitive house in the background, the house in foreground takes on rather a dramatic pose.

2

3

[アイソメ図]左下がATL8110。右上はその斜め前に建っているPEH8501(岩崎邸)で、前者が完成して4年後に設計を依頼され、2軒が連続的な関係をもつに至った。

Below left in the isometric drawing is ATL 8110. Upper right is PEH 8501 (The Iwasaki House), standing diagonally opposite to it. It was ordered four years after the completion of the earlier structure.

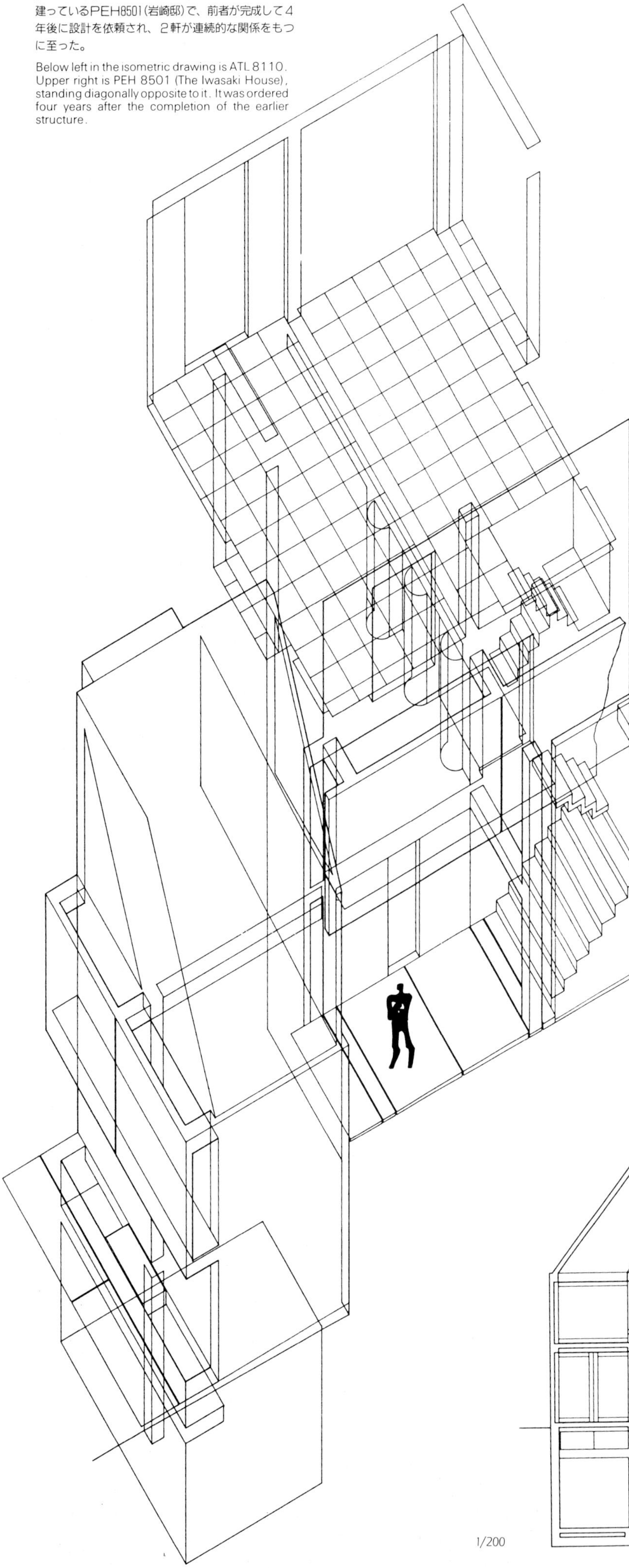

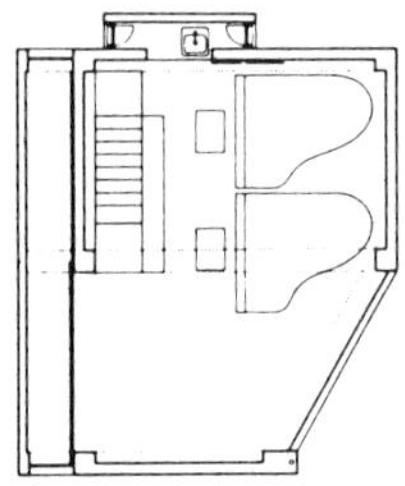

Second floor plan

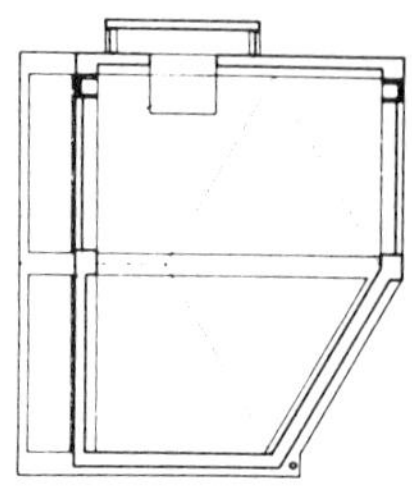

Third floor plan

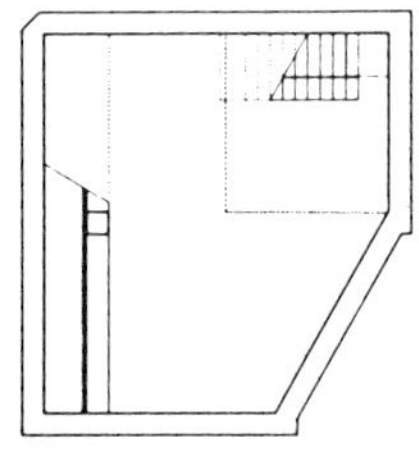

Rooftop plan

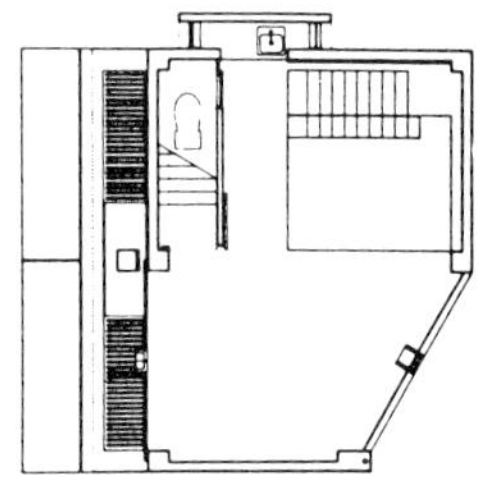

First floor plan 1/200

4

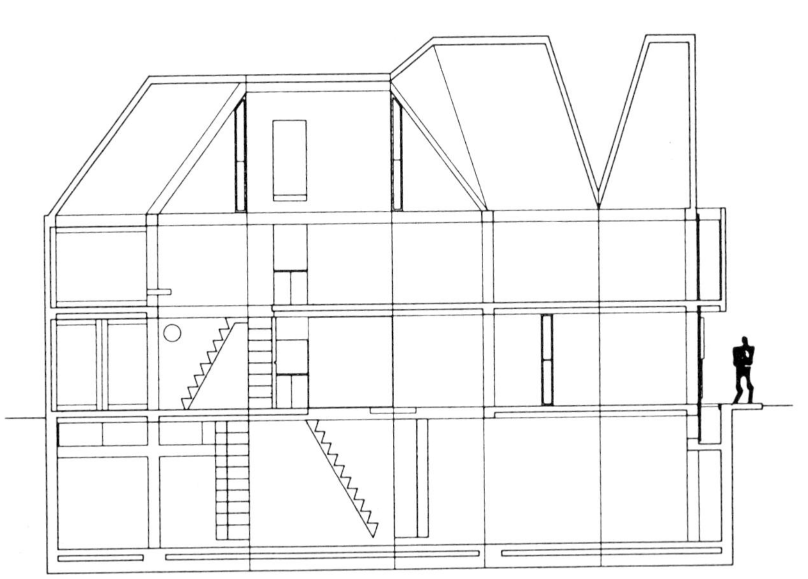

1/200

5

6

7

4 2階より階段室を通して下を見下ろす。
5 入口から1階アトリエを見る。大きなはめ殺しガラスの中央に、手動の風抜きがある。ガラスの向うに伸びた路地。
6 1階から階段室を通して2階を見上げる。階段室は普断はドアで隠されている。
7 1階から階段の吹き抜けを通して、地下のアトリエを見下ろす。

4 Looking down the staircase from the second floor.
5 Looking at the first floor of the atelier from the entrance. In the center of the glass, there is a hand-operated wood ventilator. Extended line to the other side of the glass.
6 Looking up to the second floor through the staircase. The staircase usually is hidden by the door.
7 Looking down into the basement through the stairwell.

8

9

10

8 2階。天井を見上げる。2階分の高さの壁の一部が2重になって、光の入るアルコーブをつくり出している。コーン状の空間。
9 2階。ピアノ室兼書斎。右手の壁は閉じているが、柱からわずかに外に張り出し、スカイライトをとって小さな壺庭のようなスペースをつくり出している。
10 斜線制限がそのまま屋根の形となっており、また室内の形にもなっている。

8 The second floor. Looking up at the ceiling. A part of two-story wall is doubled to create an alcove to take in light. Cone-shaped space.
9 The second floor. A piano room and study. The wall on the right is closed, but slightly extrudes from the pillar to create a small garden, which takes skylight.
10 Oblique line restrictions form the shape of the roof while also shaping the inside of the room.

鈴木了二[SUZUKI Ryoji]

杉並の住宅 物質試行13[Experiences in Material 13]

■鈴木了二の建築にはすべて、「物資試行」という名称が付けられ、それに設計年代順に通しナンバーが打たれている。端的にいえば、建築を説明したくない、より即物的につくっていきたいという決意の表明である。同時に、紙や布は物質感が弱く、鉄やコンクリートはより強いという印象は単なる先入観であり、あらゆる素材はそれぞれ同等の物質的力をもつ存在であることを具体的に示していきたいという考えに基づいている。
そのような姿勢から、彼は、一方ではチャールズ・レニー・マッキントッシュの建築に影響を受け、他方では街なかの、崩れかけたバラック建築にも興味を抱く。そこに異なる素材の対照、素材の異常な収斂を誘う様式性、つくられた建築を超えて出現する、時間や住むことによる物質の侵蝕、を見ているのである。
彼の手がける建築では、だから、シェルターや囲われた壁をつくるために動員されたさまざまな素材が、用途を超えて、素材そのものとして見える領域を求めて宙に飛び出し、異なる素材と遭遇する地点を探している。彼は建築の属性である構築を嫌い、例えば彫刻の特性である切開、掘削、破砕を信頼する。打放しコンクリートの壁は、そうした効果をもつ素材として、よく用いられる。つまり、本来は充填材である素材が、切り裂かれるべき母胎としての素材に転じている。

■建築家の説明によれば、打放しコンクリートのマッスが、内部で、全体を直交する2本のトップライトによって3つのゾーンに分割されている。道路側の、玄関・階段室・予備室、コンクリートのラーメンに収められたダイニングルーム・キッチン・浴室・アトリエ、木造の茶室と寝室のゾーン。これら3グループに分けられた部屋々々は、トップライトの下の土間や露地や通路によって、島嶼のように明快に識別できる。同時に、その境界部分に、鉄骨や角材や曲ったコンクリートの壁や障子が集中的に顔を出し、家全体に張り巡らされた不協和音的なエレメントを束ねた光景となっている。日本の街の一部を切り取った姿とみれば、この光景は、個々に閉じた家や店舗の隙間に充満する電柱や電線やストリート・ファニチャーのそれと同じであり、それらが外から静かな室内を浸しはじめているグロテスクな状況を、しかし詩的に映しとった、アンビバレンツな住宅として意識される。

■All the structures of Mr. Ryoji Suzuki have the name "Experiences in Materials", with a serial number given in order of design. This is a declaration of his unwillingness to explain about his structures, and of his willingness to construct materially. At the same time, it expresses his belief that percieved differences between iron, concrete, paper, and cloth are simply preconceptions, and all materials are actually the same.
In this view, he was influenced by the architecture of Charles Renie Mackintosh. At the same time, he is interested in collapsable barrack structures. He observes the contrast of materials, stylization comming from the abnormal convergence of materials, the ageing of materials through time, and the dwelling which appears over the structure created.
In his architecture, a variety of materials are activated to form shelters. The walls look for a point to encounter different materials. Materials jump out to create areas where they are identified as themselves beyond their purposes.
He disdains the architectural construction, and instead trusts to the forms of sculpture, such as sectioning, excavation and crushing. Ferroconcrete walls are often used as materials for this effect. This is to say that the materials to be used as fillings are transformed to be used as the mother materials to be torn up.

■According to the architect, the ferroconcrete mass is divided into three zones by two sky-lights crossing the whole structure at right angles. These are the zone of the entrance hall, staircase, and spare room; the zone of kitchen, bathroom, and atelier within a rigid concrete frame; and zone of the wooden tea room and bedroom. These three zones are clearly set apart — like islands — by the floor, and lines of the sky-lights.
At the same time, steel frames, rectangular timbers, angled concrete walls, and sliding doors are concentrated around the borders to highlight the dissonant elements in the house. If one sees it as part of a street in Japan, this is a scene of closed houses and shops, filled with electric poles, wires, and street furniture. It leads one to view the house as an ambivalent presence poetically reflecting the grotesque situations invading the quiet rooms from outside.

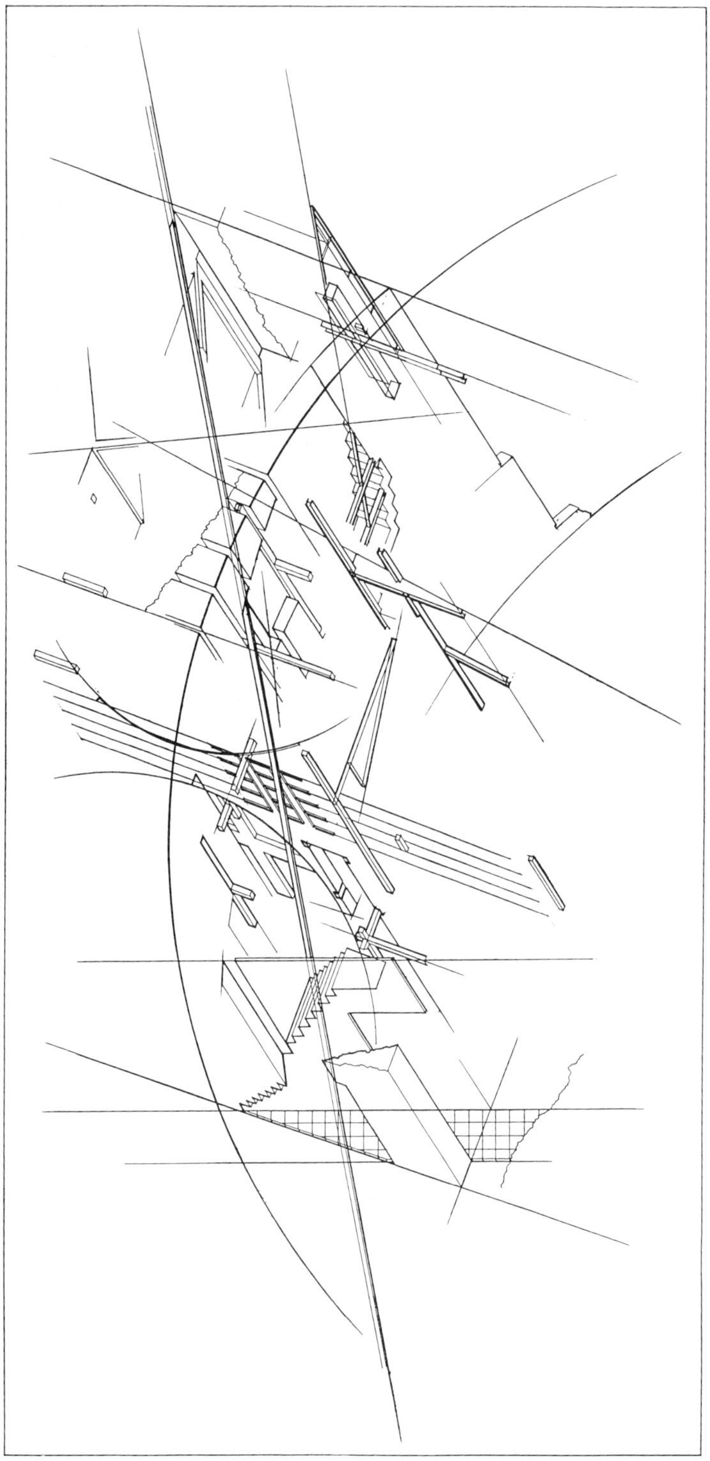

■2階アトリエから、吹き抜けを通して和室の外壁を見る。

■The outside wall of the Japanese-style room is viewed through a well from the atelier on the second floor.

1

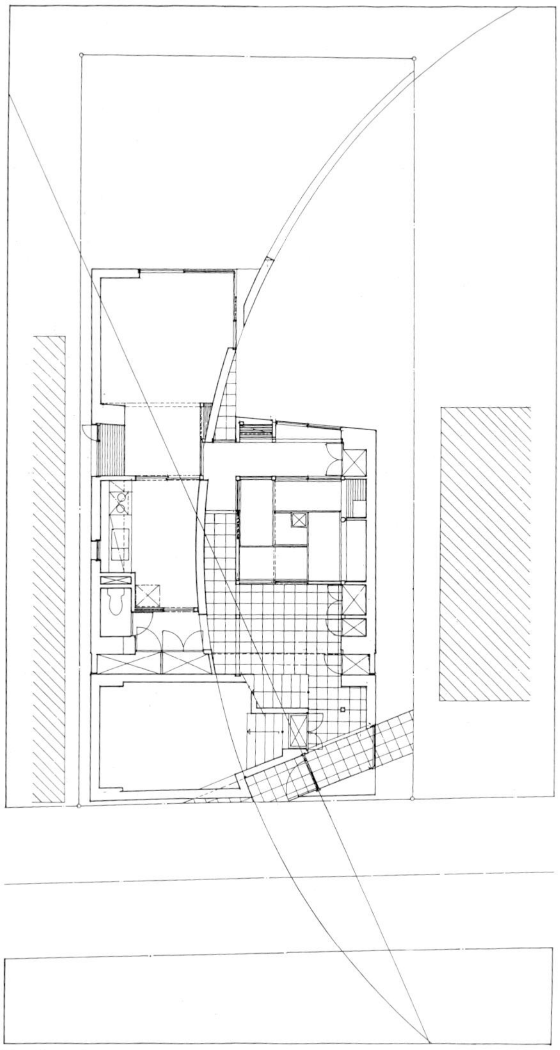

First floor plan

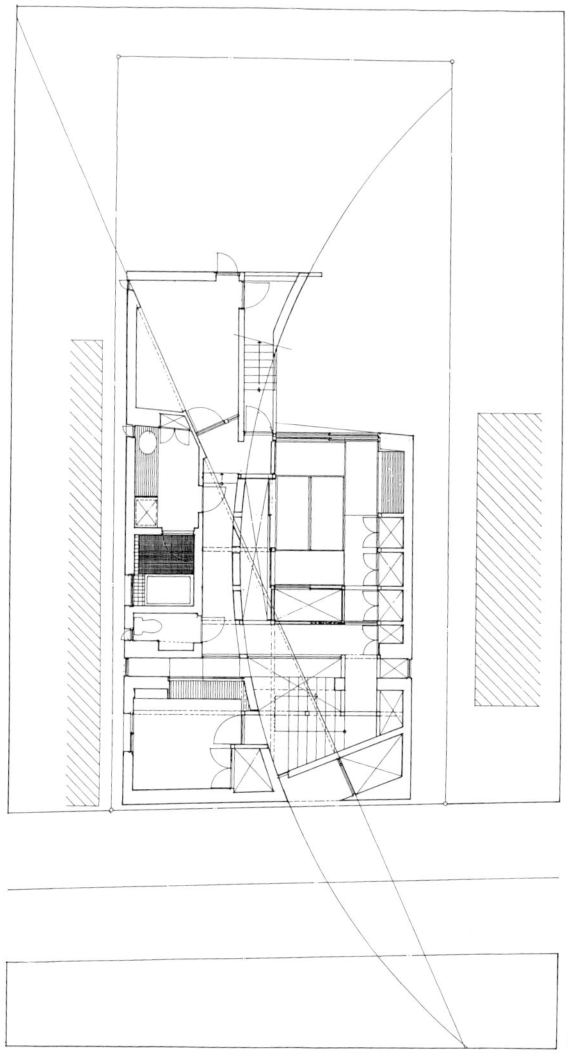

Second floor plan

1/200

2

4

3

2道路(北)側外観。右手、赤いドアが玄関。段々に切られた壁が唯一の表情。
3屋上階段の足元。赤と黒にペイントされた鉄骨の手摺。
4階段室から土間と茶室を見下ろす。
5茶室とキッチンの間の露地。奥に縁側。
6 1階。茶室内部より土間(玄関ホール)を見る。

2Appearance from the street north side. The red door at right is the entrance. The wall, with its levelled cuts is the only expression.
3The foot of stairway to the roof. Rail of iron frames painted in red and black.
4Looking down at the tea room from the staircase.
5The area between the tea room and kitchen. Verandah at the back.
6The first floor. Looking in to the entrance hall from inside the tea room.

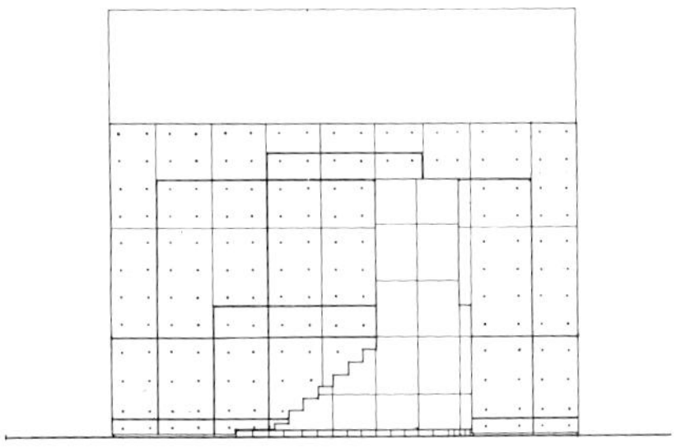

North elevation 1/200

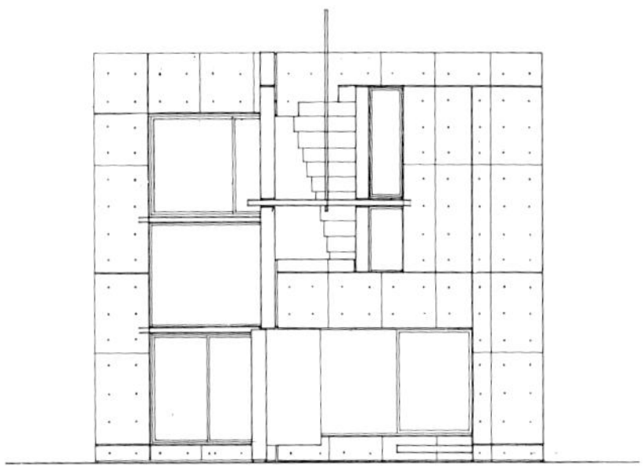

South elevation

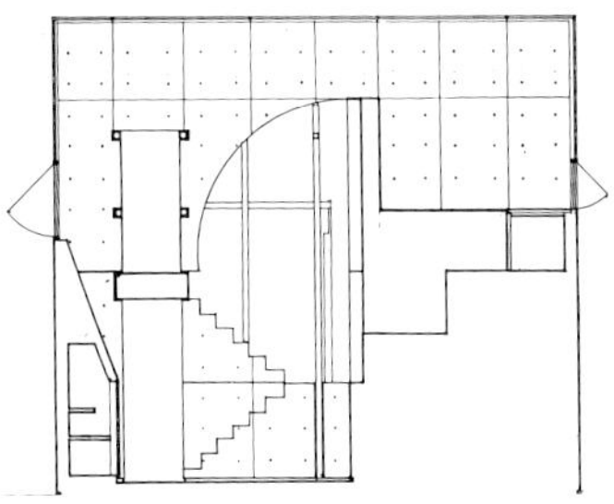

Section

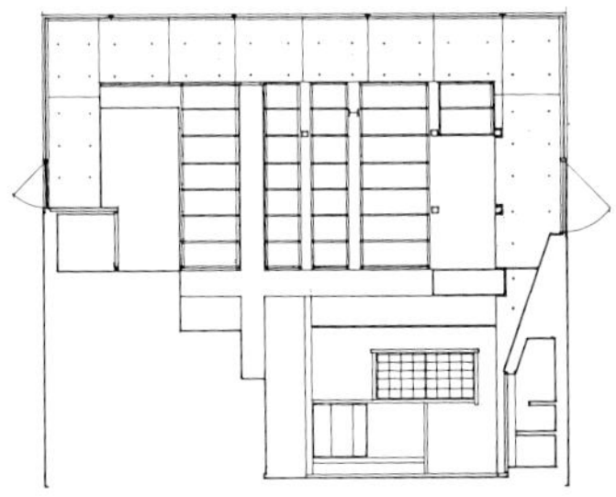

Section

5

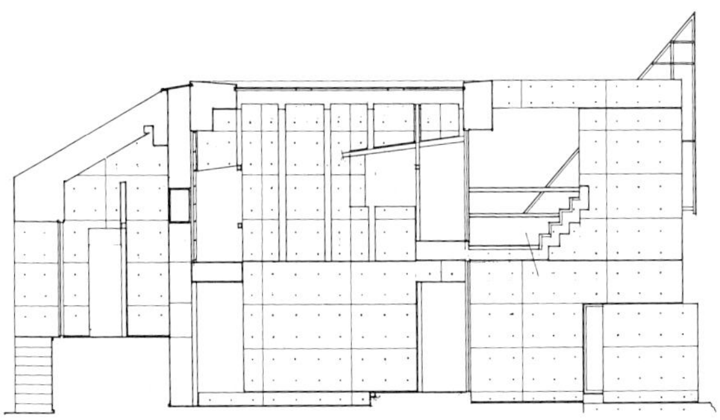

Section 1/200

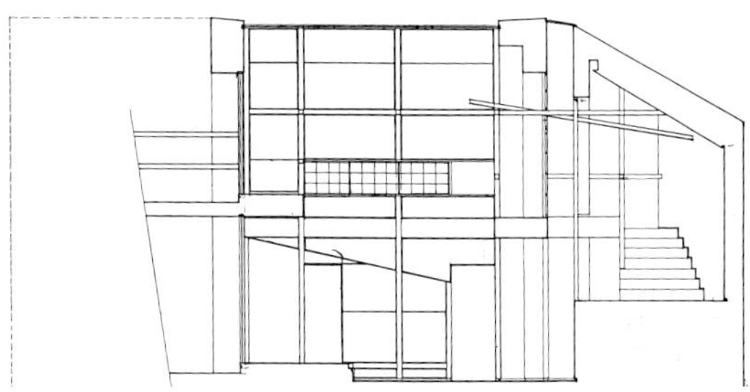

Section

6

7

9

8

7 2階。洗面・浴室前の通路から天井を見上げる。円弧のコンクリート壁にスリットが開けられて、その上部の天井部分が切り抜かれている。
8 土間から階段室を見上げる。玄関ホールの空間に剥き出しになった階段。
9 2階。寝室の壁は、スリット状の吹き抜けに面している。
10 1階。茶室とキッチンの間の露地の上は吹き抜けて、スカイライトがかかる。半ば外部空間としてのスリット。

7 The second floor. Looking up at the ceiling from the passage in front of the bathroom. A slit is made in the concrete, cutting the ceiling above.
8 Looking up the staircase from the ground floor. Bare stairs in the entrance hall.
9 The second floor. The wall of the bedroom is facing the sky-lit well.
10 The first floor. The area between the tea room and kitchen is a well lit with skylight. A slit partly as an outer space.

10

11

12

11 2階寝室内部より、露地の吹き抜けを見る。
12 1階茶室と2階道路が、玄関ホールの吹き抜けに面している。その空間の大きさを縁取るような円弧のコンクリート壁。
13 2階寝室前の通路。東から西の方を見る。浴室前の通路と対照的な透明性。
14 階段側から寝室前の通路を見る。
15 庭(南・西)側にまわると、外壁の各エレメントのおしゃべりは活発になる。

11 A view of the well on the alley, from inside the bedroom on the second floor.
12 Tea room on the first floor and the second floor passage face the well of the entrance hall. A concrete arc wall rims the space.
13 Passage in front of the second floor bedroom. Looking from east to west. It is transparent in contrast to the passage in front of the bathroom.
14 Looking at the passage in front of the bedroom from the side of stairs.
15 Turning to the garden side (southwest), the elements become more expressive.

13

14

15

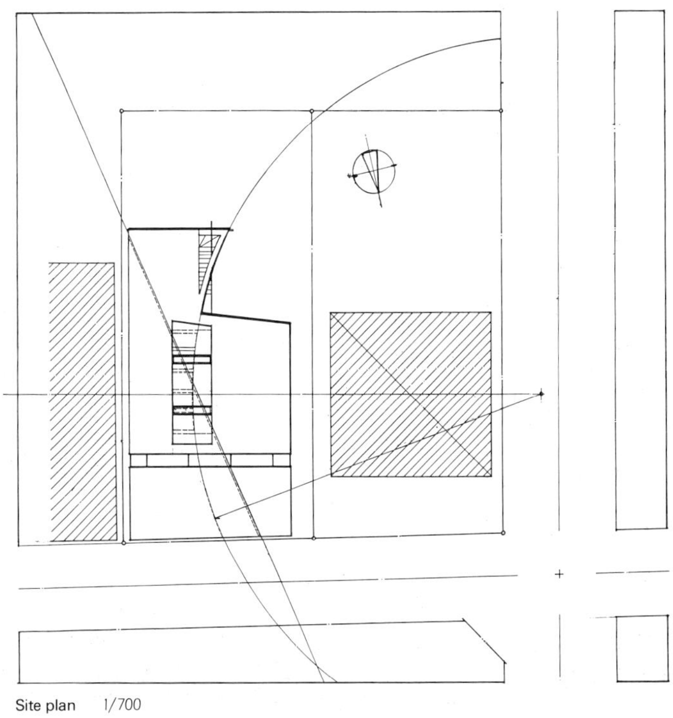

Site plan　1/700

A商店ビル 物質試行21[Experiences in Material 21]

■小中規模のオフィスや店舗のビルが建ち並ぶそのひとつとして、1階にガレージ、2階にオフィスを置き、3階から5階を住居としている。アルミニウム、鉄、ガラス、コンクリートに加えて、第2の素材と呼べるような、強烈な赤と緑の色彩が、相互に象嵌し合っているかにみえる、つまり、素材そのものと表面とがそれぞれ自立し等質になっている、きわめて精度の高い建築である。
杉並の家においてもそうだったように、鈴木はここでも、直線や円弧によって建物のマッスを切開する手続きをとっている。しかしその「法則」はこれまで、平面において、建物全体をいくつかのゾーンに切り分ける距離の長さで行われていた。それに対して、この住居においては、例えば円弧は、建物の端部を膨張させ、あるいは浴室・便所のようなコンパクトな部屋の外の輪郭を見せるような、配置や小規模のスケールで使われている。さらに特徴的なのは、大きな円弧の線が断面方向において見え隠れしながら大がかりに用いられていることである。切り裂かれるためのコンクリートや金属の面、つまり切り裂かれてさらに剝き出しになっていく素材の「身体」は、この新しい法則の設定によって、何か身籠るような表情に転じてきている。素材の力に依存しがちだった「物質」から、空間の全体性への展開が感じられるのである。

1

■This fits in with the other buildings of offices and shops standing with it. There is a garage on the first floor, offices on the second floor, and a house from the third to the fifth floor. This is the architecture of extremely high precision, using aluminium, iron, glass, and concrete, together with dazzling colors of red and green. These colors can almost be called secondary materials, they look as though they were inlaid with each other.
As with the Experiences in Material 13, Mr. Suzuki, here cuts off the mass of the building with straight lines and arcs. This idea has been executed over the dimensions of the building by dividing it into several zones. The ideas of that house are seen here over a smaller scale, with the outlines of the compact rooms — such as bathroom and toilet — expanding the rim of the building. More characteristic is the line of arc appearing and disappearing in sections. The face of the concrete or of the torn metal, that is, the "body" of the material is, by the execution of this idea, transformed from "material" to living body.

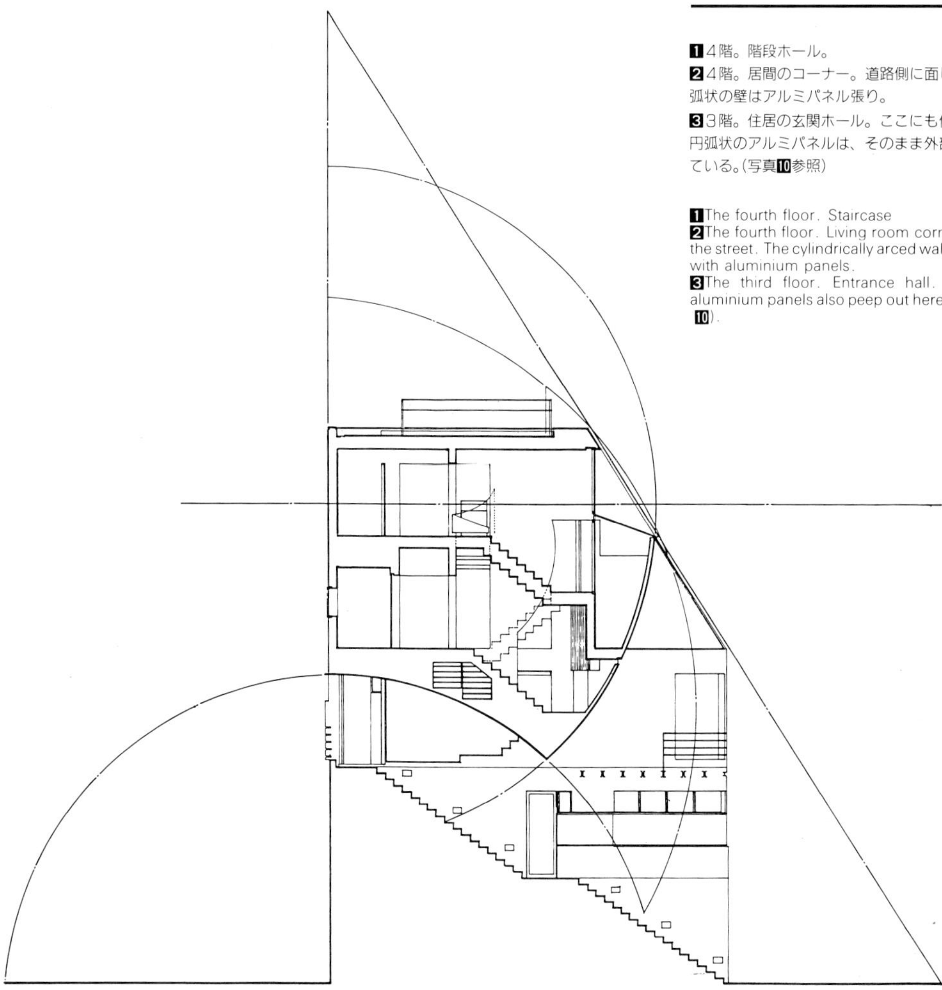

1 4階。階段ホール。
2 4階。居間のコーナー。道路側に面した部分。円弧状の壁はアルミパネル張り。
3 3階。住居の玄関ホール。ここにも使われている円弧状のアルミパネルは、そのまま外部に顔を出している。(写真10参照)

1 The fourth floor. Staircase
2 The fourth floor. Living room corner. Facing the street. The cylindrically arced wall is covered with aluminium panels.
3 The third floor. Entrance hall. Cylindrical aluminium panels also peep out here (see Photo 10).

2

3

4

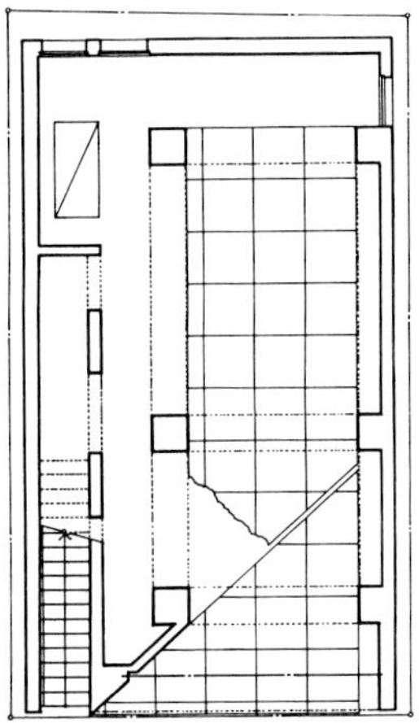

First floor plan 1/200

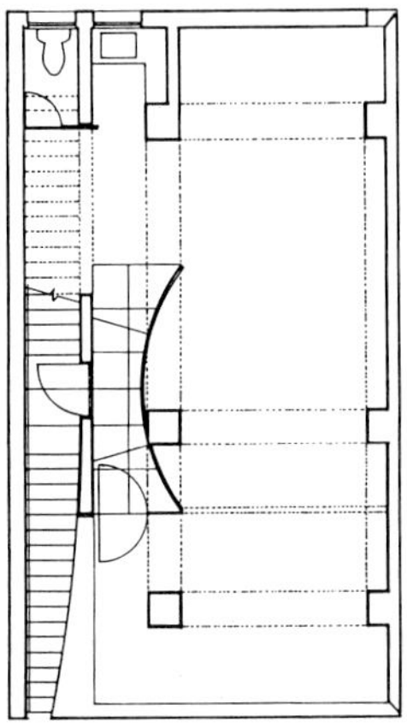

Second floor plan

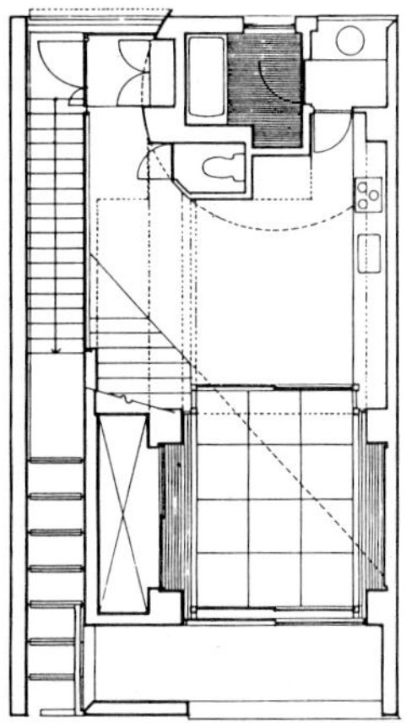

Third floor plan

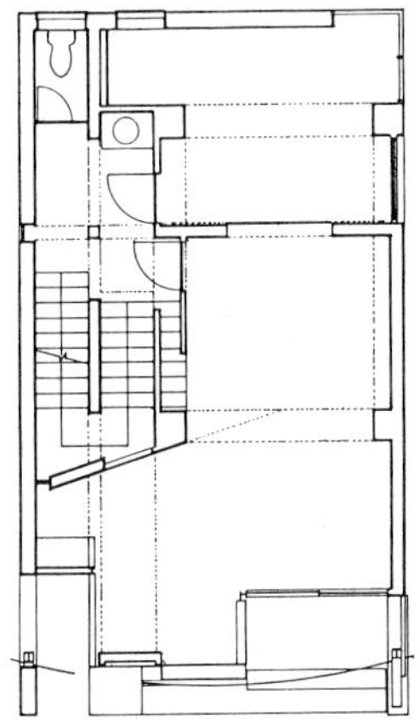

Fourth floor plan

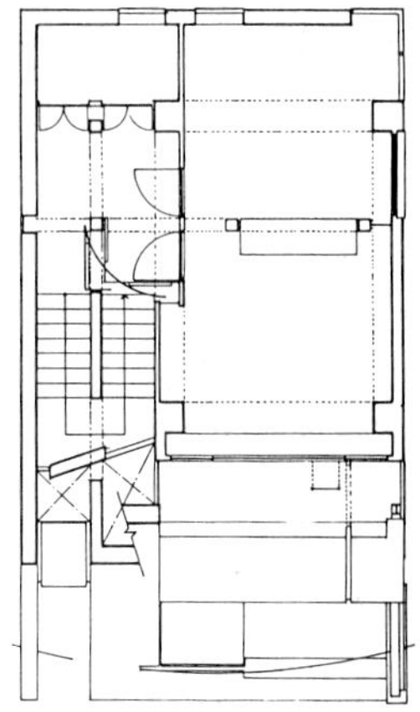

Fifth floor plan

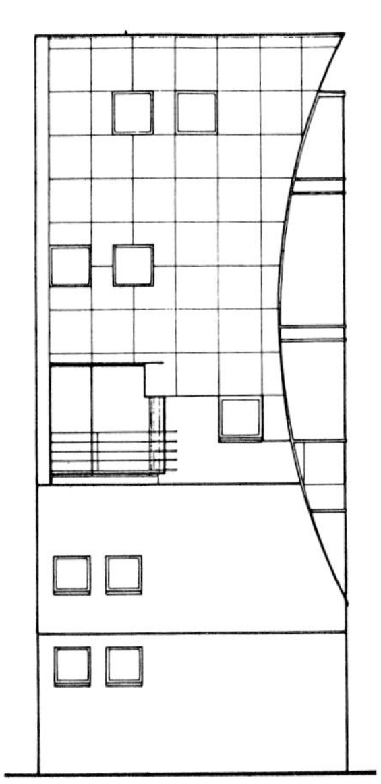

Elevation

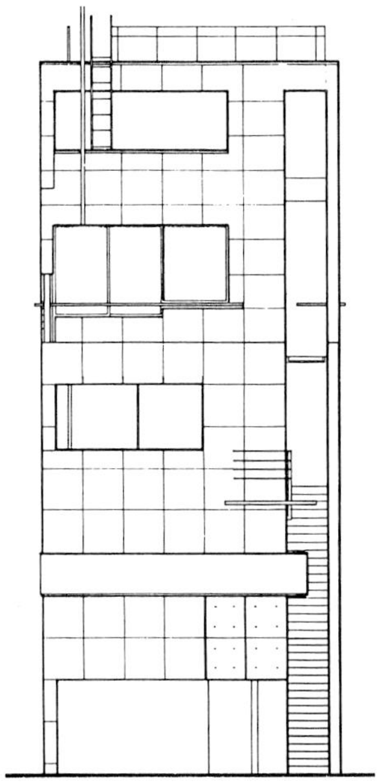

Elevation 1/200

5

6

4 4階居間。
5 5階。子供部屋前のホール。三角形のアルミパネルを円弧状に曲げて、手摺にしている。
6 4階。居間から寝室を見る。居間との仕切りは、グリーンに塗られた引き戸。

4 The fourth floor. Living room.
5 The fifth floor. The hall in front of the children's room. Triangular aluminium panels are bent in a circular arc and used as balusters.
6 The fourth floor. Looking at the bedroom from the living room. The living room is partitioned by green sliding doors.

7

7 1階から3階まで、一直線に階段が貫いている。左側の独立壁を赤く塗られた鉄骨が支えている。写真でいちばん上部(道路側)のH型鋼だけは化粧。グリーンに塗られている。
8 9 道路側外観詳細。ステンレスの手摺、グリーンに塗られたH型鋼、ガラス等が、道路側から階段室上の吹き抜けに関わっている。

10 裏側外観詳細。玄関ホールのアルミパネル壁が、ステンレスの手摺にからまって外に突き出ている。
11 道路側外観。コンクリートと鉄、ガラスとの精度の高い取り合わせ。
12 裏側外観。

7 Straight stairway penetrating from the first floor to the third floor. The independent wall on the left is supported by a red-steel frame. Only the H-beam at the street side is painted green in the photograph.
8 9 Detailed appearance from the street side. Stainless steel balusters. Green H-beams and glass are involved with the well above the staircase (from the street side).

10 Detail on the rear side. The aluminium panelled wall in the entrance hall clings to the stainless steel balusters and extrudes outward.
11 Appearance of the street side. A highly precise arrangement of concrete, steel and glass.
12 Appearance of the rear side.

8

9

10

11

12

高松 伸[TAKAMATSU Shin]

修学院の家Ⅱ[House in Shugakuin Part Ⅱ]

■高松伸の初期の仕事でもっとも高名な「織陣」と、最近作でこれもまたひとつのピークを形成した「キリンプラザ」に見られるように、彼のファサードの多くは、石と金属で隙間もなく充填されている。そして極度に私的な古典主義といえるようなシンメトリー。また高松自身が「小さな宝石箱をそのまま大きくしたような」という、人間の手がかりを振り払うようなスケールの爆発的肥大。
この特性は、打放しコンクリートの小規模な、例えば住宅建築においても全く変わらない。コンクリートの裸の肌は被覆されることを待ち受けていない。内側には石や金属と同質の血が巡っていて、いつだって狼男になれるのである。
饒舌な装飾や形態によって、彼の建築は逆に沈黙し、どんな応答をも拒否しているように見える。しかし高松は決して神話的建築家ではない。むしろもっとも身近かに感じられる同時代作家である。
彼の建築は、その場所に馴染まず、いつか離陸するような不安感をもつ。しかし、それこそ高松の建築の揺ぎない根強さの証拠である。自分の建築のなかに、高松は体験してきた建築の前衛の歴史を映し、また引き継ごうとしている。
例えば、とくに住宅においてはその見事な、練りに練ったプランニングは、外観にも確信となって現れている。だから彼の装飾的なファサードほど内的必然性をもつ例は他に少ない。仮面ではないのだ。

■京都の緑ゆたかな住宅街にある。高松の住宅建築の多くが、宝石箱のようにぴっちり閉じているなかで、この住宅は、不思議な印象で道路に開いている。
東側は長い直線の、西側は円弧状の壁で貝のように合わさっている。その隙間から円弧の壁に沿って玄関に入ると、そこはすでに家の中心部であり、玄関ホールは即、大階段となって、半ば屋内半ば屋外に立ち上がっている。階段を上がりきって、引き返すように子供室の前の廊下を通り、玄関ホールの真上をブリッジで渡り、そのまま屋外に出ると、南の端にある四阿風のテラスまで、ブリッジは一直線に伸びている。いわば家の奥へ奥へと入り込んでいく気持ちなのに、もとの道路ぎわに出てしまう。ル・コルビュジエの「建築的散策」を思い出させるような動線だが、この独特な内外にわたる通路が、家全体の部屋や庭の性格を決定する骨格となっている。
例えば、1階の南半分を占める居間は分断されて、中央に中庭が割って入る。2つに分かれた居間は、階段とテラスと中庭の圧力を受けて半ば地下に沈み、隠されたパブリック・スペースとなっている。
北半分においては、上下階とも、長い通路の延長のごとく、中廊下式のプランが個々の部屋を分離・独立させて、落ちついた領域をつくり出す。
子供部屋まわりの壁は上方に雲のように開いて、水滴形の断面の照明塔が天から与えられる。遠方から投げ入れられた円弧の断片が、家全体を少しずつ押し流し、また天へ、地下へと生長させている。

■His famous "Orijin" from his early work, and his latest work "Kirin Plaza", show facades heavy with stones and metals, which evidence a symmetry which may be called an extremely private classicism. An explosive growth of scale, as if to reject human hand, is said by Mr. Takamatsu himself to be "like a small jewel box enlarged as is".
This characteristics is exactly the same for his small-scale ferro-concrete structures like houses. The bare skin of the concrete does not expect to be covered. Inside of it is the same blood as the stones and metals, and it can become an wolf man anytime.
By eloquent decoration and style, his architecture appears silent, refusing to give any answers. Nevertheless, Mr. Takamatsu is not a mythical architect. Rather, he is a real and familiar achitect of our age.
His architecture has a feeling of uneasiness — as if it were out of time and place. This is proof of intensity of Mr. Takamatsu's architecture. Mr. Takamatsu is continuing the history of vanguard architecture in his own architecture. For instance, especially with houses, extraordinarily elaborate planning indicates confidence in its appearance. There are few examples with decorative facades. He does not believe in disguise.

■This house is located in the green-filled residential district of Kyoto. Although most of Mr. Takamatsu's architecture is closed like a jewel box, this house is open to the street,

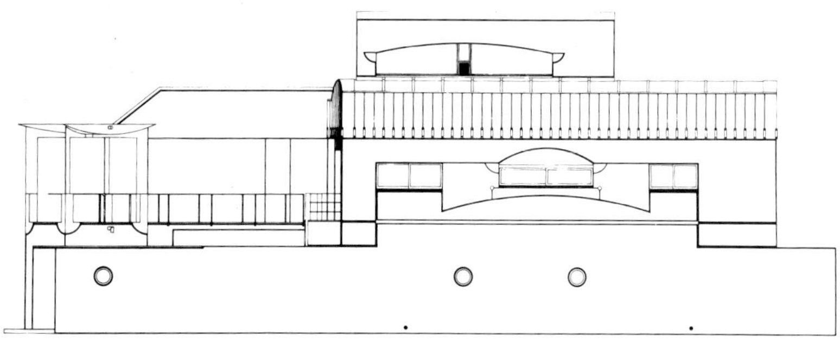

East elevation 1/250

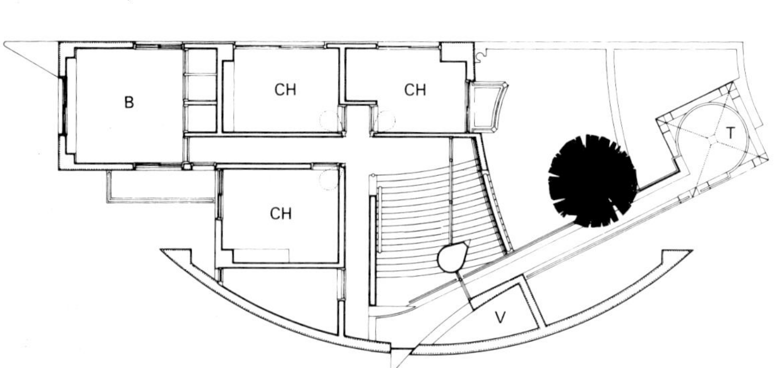

Second floor plan

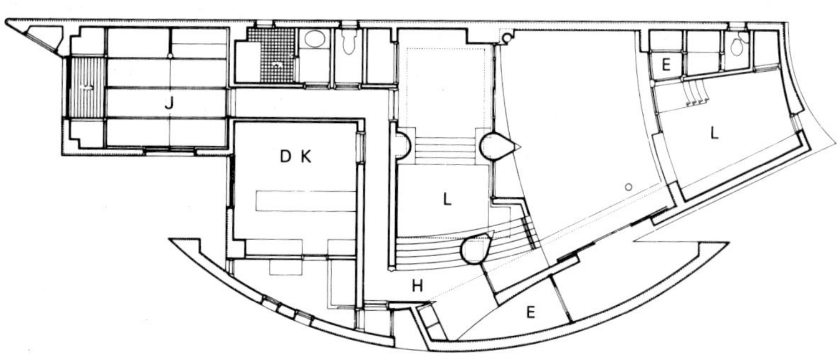

First floor plan 1/250

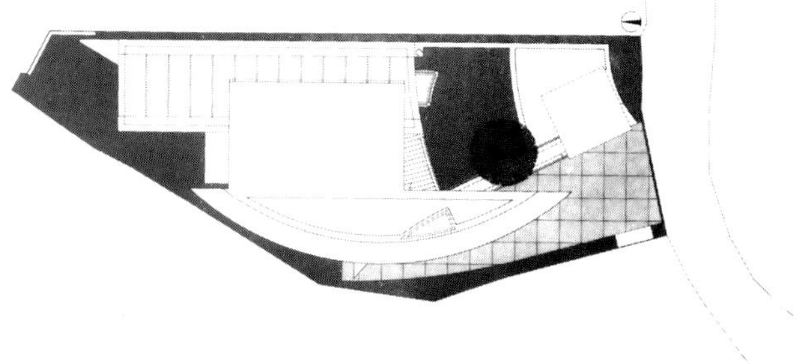

Site plan 1/400

creating a strange impression.
It is united like shells with a long straight line on the east side, and an arc wall on the west side. When you enter from the slit along the arc wall to the entrance, you find yourself in the center of the house, and the entrance hall naturally becomes a large stairway leading up half inside and half outside. Going to the top of the stairs, then walking back through the corridor in front of the children's room, you go outside passing over the balcony above the entrance hall. This balcony extends straight to the terrace at the southern end.
You feel as if you are going deeper inside the house, but eventually, you'll come out to the original street side. It is a path which may reminds you of the "architectural wanderings" of Le Corbusier. This peculiar passage in and out sets the character of the entire house and gardens.
For instance, the living room occupying the southern half of the first floor is divided to have a courtyard in the center. The two living rooms are half sunk by the pressure of the stairway, terrace and courtyard, to form a hidden public space. In the northern half, the center-corridor plan separates each room into a quiet, independent area. The wall around the children's room opens upward like a cloud, opening into the lighting tower like a waterfall. Segments of the circle thrown in from the distance are pushing the house little by little to make it grow to the heaven and to the underground.

1

1 道路(南)側ファサード。道路を隔てた向かいの敷地が見える。左手に玄関。2階テラスが前面に突き出している。

1 Street side facade. The plot on the other side of the street can be seen. Entrance of the left. The terrace of the second floor protrudes.

2 3 曲面のコンクリートの壁に包み込まれるようなアプローチと玄関。
4 玄関からアプローチ方向を見る。
5 アプローチから上を見上げる。
6 玄関を入ったところで、空間はさらに高く、狭くなる。
7 玄関ホールから、アプローチ、道路方向を見る。

2 3 Approach and entrance surrounded by the curved wall surface.
4 Looking at the approach from the entrance.
5 Looking up from the approach.
6 Inside the entrance, the space becomes high and narrow.
7 Looking in the direction of the approach and the street from the entrance hall.

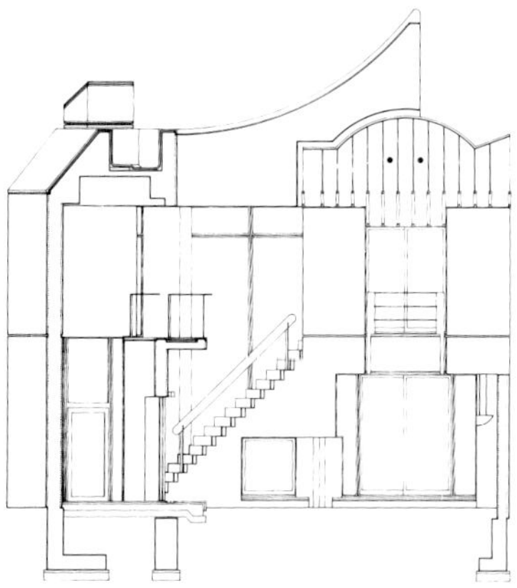

Section 1/200

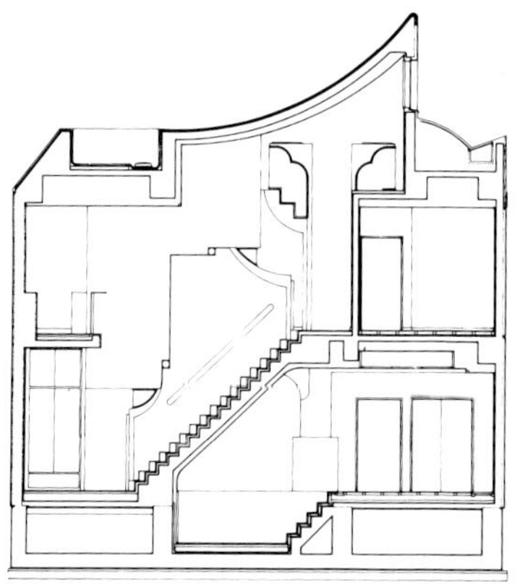

Section

2

3

4

5

6

7

8

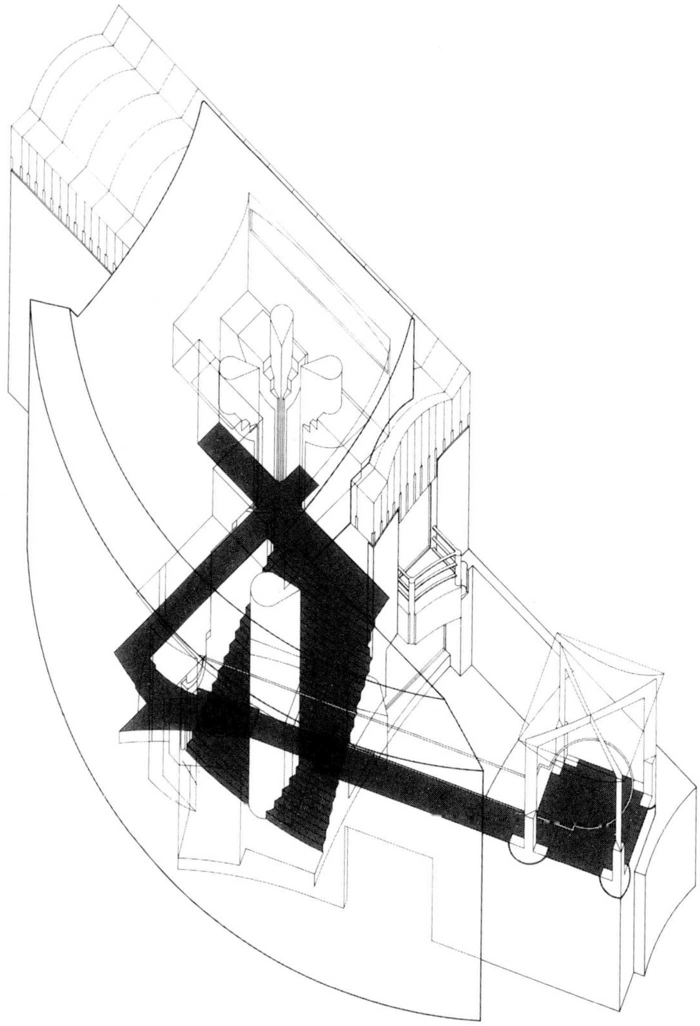

9

10

11

8 階段室から中庭を見る。階段の南側は屋外に露出している。階段室の下に居間Aが隠されている。
9 南端のテラスとブリッジ。
10 テラスから見る屋根のシルエット。

11 中庭に面した南側ファサードを見上げる。右側は子供室とバルコニー。

8 Looking at the courtyard from the staircase. The south side of stairway is exposed to the outside. Living room A is hidden under the staircase.
9 The terrace and bridge on the south end.

10 Silhouette of the roof as seen from the terrace.
11 Looking up at south facade facing the garden.

12

13

14

15

12階段室から2階個室まわりの壁を見上げる。
13玄関ホールと階段室の上部にかかるブリッジ。左手はそのまま屋外に。
14中庭の上部を渡るブリッジから、屋内外にわたる大階段を見る。
15個室まわりの壁と天井。
16階段から2階廊下、腰壁への連続。

12Looking up at the wall around the independent rooms of the second floor from the stairway.
13The balcony above the entrance hall and the staircase. The left side leads directly outside.
14Looking at the large stairs inside and outside of the house from the balcony crossing above the courtyard.
15Wall and ceiling around the independent rooms.
16Continuation from the stairs to the corridor and the basewall of the second floor.

16

17

18

17 2階。個室まわりの壁は、コーナーで照明塔を抱きかかえるようなデザイン。
18 1階和室の西面。
19 ブリッジを通して南端のテラスを見る。

17 The second floor. The wall around the independent rooms is designed to hold the lighting tower.
18 West side of the Japanese room on the first floor.
19 Looking at the south terrace through the balcony.

19

1

■京都の市心ではあるが、新しく開発された住宅街の一画。間口が広く、奥行の浅い敷地に建つ。道路の際までコンクリートの長い壁が迫り、南と北に庭を取り込み、建物と一体化している。
1階から3階まで、エレヴェーションもプランもほぼシンメトリーにまとめられており、各階の表情を変えて3層構成を明快にしている。1階は無窓の壁が、ガレージと倉庫、トレーニング・ルーム、そして南北の庭を囲う。2階は御影石の装飾に異形の窓が並ぶ壁。居間、食堂、台所のゾーンに、階段室を挟んで書斎のゾーンが相対する。その上は、両端を斜めにカットした半円形の鉄骨を、スチール板で覆った寝室階。ヴィツェンツァのパラディオのバシリカを思い出させるような古典主義的な様式を匂わせている。
装飾的要素は、ほとんど1ヵ所、中央部分に集中する。ミラーガラスの楔形は、外側にはその底部に照明器具を結晶のようにつくり出し、内側に向かっては、石を貫いて玄関ホールに顔を出し、2階の階段室踊り場から突き出したゴンドラ風のバルコニーとなって同じ形を焼き付ける。このバルコニーには書庫わきのアルコーブと、反対側の階段室が相対して、建物の中央を東西方向に貫通する。同時に玄関ホールの吹き抜けは3階にまで達する。
中央部に、建物への腐食作用をもつ装飾、あとは周縁部に湧出する気配としての装飾――赤く塗られた太いパイプ。

■Though at the center of Kyoto, in a new residential area. This house stands on a plot of land with a wide frontage but little depth. A long concrete wall comes up to the edge of the street. The gardens on the south and north are incorporated as parts of the structure.
The elevation and the plan are designed to be more or less symmetrical, but by giving different expressions to each floor, the three storied composition is explicit. On the first floor, a non-windowed wall surrounds the garage, storage room, training room and the gardens on the north and the south. On the second floor, the wall has a series of uniquely shaped windows with granite decorations. The staircase is in the center, the living room, dining room, and the kitchen zone, on one side, the study on the other. On the floor above, the bed room floor has semi-circular steel frame which is covered with steel plate and cut slanting along the western edge. This reminds us of the Basilica in Vizenza, and seems inclined to classicism on the whole.
The decorative elements are all concentrated almost in one place, the center. Wedge-shaped mirror glass has the illuminates like chrystals at the bottom of the outside, and shows through the stone on the inside, then continues on from the gondola-shaped balcony, sticking out of the second floor staircase, giving a strong impression of regular shapes.
This balcony runs through the center of the building from the east to the west. It has an alcove at the side of the library, and the staircase on the opposite side, facing each other. The stairwell from the entrance hall goes up to the third floor.

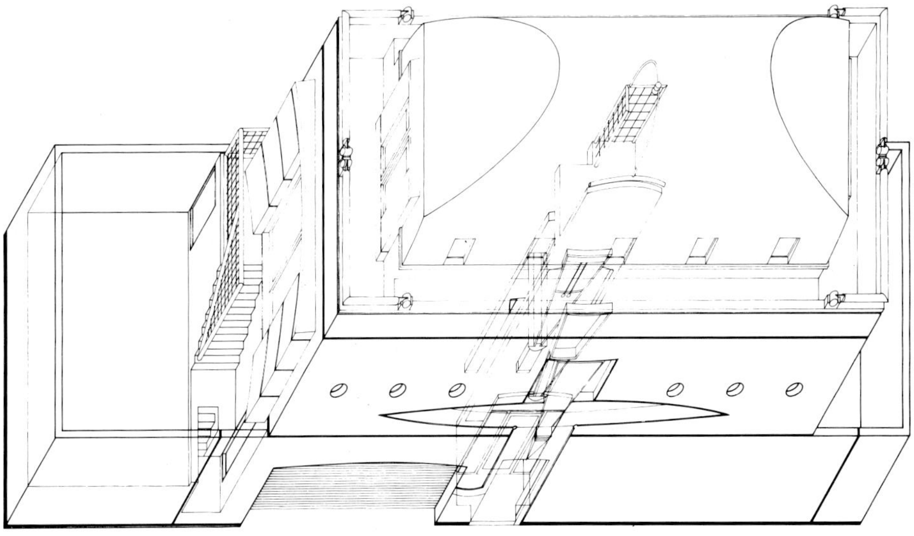

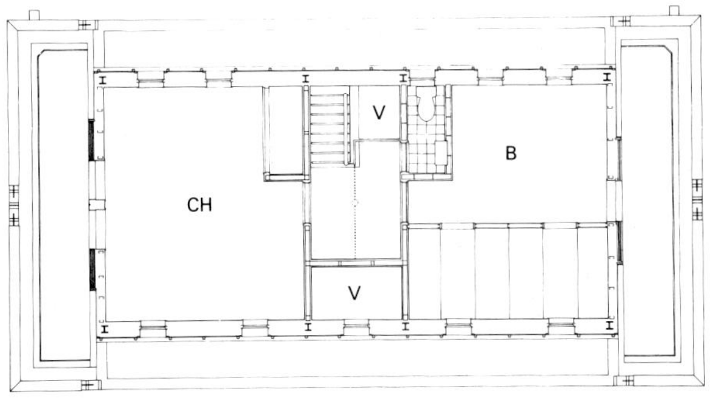

Third floor plan

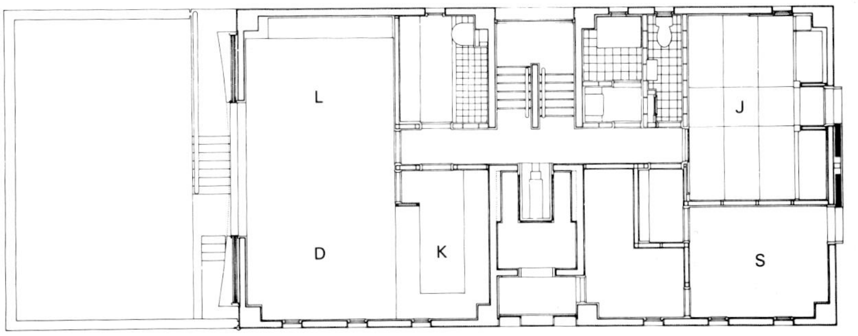

Second floor plan

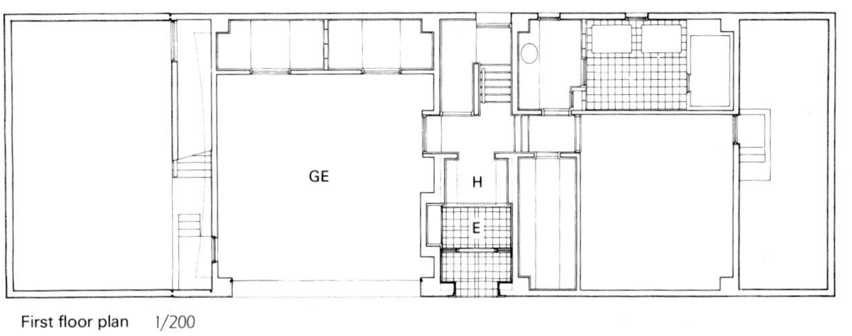

First floor plan 1/200

1道路(東)側玄関まわり。打放しコンクリート、御影石、スチール・サッシュの構成。

1Outside the entrance on the street (east side). A composition with ferroconcrete, granite and steel sashes.

2

3

4

2玄関のシンメトリカルな構成と、ガレージのシンメトリカルな枠との取り合い。その境にもうひとつ新しいヴォキャブラリーが発生している。
3道路側の開口部は、この丸窓が6つ、玄関の左右に並ぶだけ。屋上の手摺は赤く塗られたパイプ。
4奥に浅く、間口の広い敷地から決定されたファサード・デザイン。

2The symmetrical composition of the entrance, and the symmetrical garage-matching well. In between them, at the border, another new vocabulary is originated.
3Only these 6 round windows are placed in rows on either side of the entrance open on to the street. The handrail at the top of the roof is a pipe painted in red.
4The plot with little depth and a wide frontage determined the facade design.

5

5 玄関からホールを見る。磨かれた御影石が鏡のように空を映す。その石に楔のように打ち込まれた照明灯。
6 玄関ホールを見上げる。写真上部が外側。玄関の真上にある逆三角形の開口部の形が内側で繰り返されている。
7 階段室から玄関ホール上の吹き抜けを垣間見る。吹き抜けには、ゴンドラ型の小さなバルコニーが突き出ている。
8 ゴンドラ型のバルコニー。底部に振子鎌のような断面の照明器具がついている。
9 和室。右手に床の間。
10 和室。床の間。

5 A view of the hall from the entrance. The well polished granite reflects the room like a mirror. The lighting is stuck into the stone like wedges.
6 Looking up at the entrance hall. The top of the picture is outside. The inverted triangle shape of the opening is repeated on the inside.
7 A view from the staircase above the entry hall to the stairwell. A small gondola-shaped balcony sticks out into the stairwell.
8 The gondola-shaped balcony. At the bottom, lighting equipment with guillotine cross-section is installed.
9 A Japanese room. An alcove on the right.
10 A Japanese room. An alcove.

6

8

7

9

10

富永 譲 [TOMINAGA Yuzuru]

魚志ん [Uoshin]

■富永譲は、都市の風景に魅せられている。もっとも普通で何気ない表情の建物をいつも眺めている。鉄骨とスレートで組み立てられた工場や駅舎や仮設に近い家々。だから看板や賑やかな表皮ではなく、その裏側にある生の建築を透視して見えてくる風景を愛しているのである。その環境のなかで、何気なく目立たずに、という建築のつくり方を、日本のとくに住宅設計を中心にしている建築家は、ひとつの理想として考える向きがある。そのなかで個性的な、一貫した思想と形を維持し展開していくのは容易ではなく、富永の建築に見られる優しさと繊細さは、彼の感性である以上に粘り強い方法論として見るべきだろう。
彼の建築は、空間のヴォリームを獲得する前に、面とエレメントの集合体であろうとする。さらには線の稠密体であろうとする。濾過器のように都市の風景を純化して、家のなかに取り込もうとする気配がある。彼の住宅に木造や鉄骨造が多いのも頷けるが、コンクリート造に対しても同じ方法がとられ、そこに独特の「漉き方」が見られる。
初期モダニズムの建築も、生の建築であり、富永はそこにも深い関心を寄せつつ、都市の風景との交点に現代の家を建てようとしている。

■湯島天神のすぐ近く。箱庭のような愛らしく古い町並みが、ほんの一握りだが、まだ残されている一画で、半世紀以上経つ木造3階建ての家々が、1軒また1軒とペンシル・ビルに建て替わりつつある。そのなかの店舗併用住宅で、間口約5.3m、奥行約12mの敷地。1、2階が関西料理の店で、打放しコンクリートを主体としながら、静かで優しい、昔ながらの雰囲気をよく再現している。
その上に住居がある。目立たぬように配慮された玄関の引戸を開けて階段をのぼりつめると、小さなバルコニーごしに裏側の家並が迫っている。そこにもうひとつ大きな引戸があって、住居の入口になる。3階に台所と食堂・居間、4階に個室、5階はペントハウスの書斎と和室。
こうした厳しい制約のもとで、必要十分な住まいを構成し、周囲の環境のなかでも控え目な姿のままに、というような課題を与えられると、富永の建築はとりわけ魅力を発揮する。打放しコンクリートや椙ベニアを主調とする、倹(つま)しく清潔な室内。部屋を区切りながら視覚的には連続させていく細部の工夫。近隣とぴったり身を寄せ合いながら、空に向かって深呼吸するようなペントハウス。西の端部はわずかに開いて、立体的な小さな裏庭を形成する。日本の下町の暮らしの形。

■Mr. Yuzuru Tominaga has always been enchanted by the urban sceneries. He looks at buildings with ordinary and casual expressions. Factories and stations, built with steel frames and slate, and houses that look temporary. So he loves penetrate the "raw" structure behind the signboard and the busy surface.
Along with this idea, a casual and inconspicuous way of building things is one of the ideals of the architect, who has been working in the designing of Japanese residences. It's not easy to be individual, and to maintain and develop the consistent ideas and forms. The sympathy and delicacy shown on Mr. Tominaga's buildings should be taken as the firm methodology rather than just his sensitivity.
His structure are inclined to be aggregates of planes and other elements before they have obtained volume and space. The lines are intended to be dense. He seems to try to take in urban scenery by refining it through a filter. It is natural that the most of his structures are of wood or steel, but even in concrete structures, the same approach is taken, and there, a very original feeling can be seen.
The structures of early modernism were also "raw" constructions, Mr. Tominaga, deeply interested in them, wants to build modern houses at the cross-roads of urban scenes.

■Near Yushima Tenjin Shrine, in a section of a town where a lovely, ancient atmosphere has been kept like a miniature garden, three storied wooden structures, built over half a century ago, have been carefully rebuilt one by one.
In the midst, there is a store-residence building, with a frontage of about 5.3m, and a depth of 12m. The first and second floors are a Kansai-cuisine Japanese restaurant using ferroconcrete as the main material, blending with the quiet, traditional atmosphere.
Above is the residence, opening the sliding door carefully, as though not to be noticed, one goes up the top of the stairs. The row of houses at the back is very close on the other side of a small balcony. There, one finds another large sliding door, the entrance to the residence. The dining room and the kitchen are on the third floor, private rooms on the fourth, and the fifth floor includes a penthouse-study and a Japanese room.
Mr. Tominaga's structures are really attractive, when given a task with strict restrictions, requiring him to compose sufficient space for a residence, and to blend with the modest appearance of the surroundings. There is a frugal and clear atmosphere, indoors, with ferroconcrete and veneer as the main materials. Using sections the rooms continue visually, although placed closed to the neighbourhood. And the penthouse seems to take a breath out of the sky. The west end opens up a little to form a garden. One style of living in a Japanese downtown.

1

2

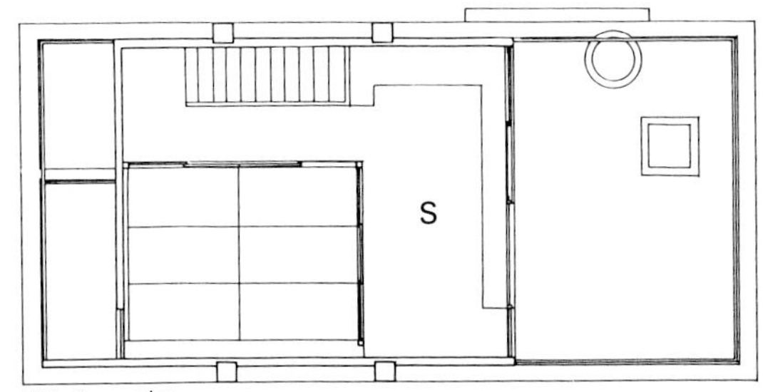

Fifth floor plan

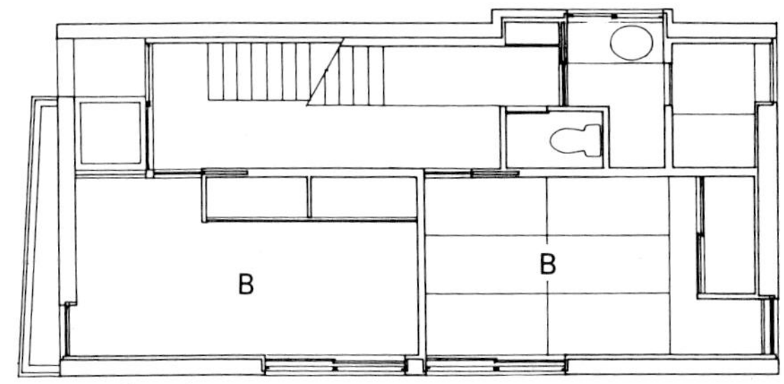

Fourth floor plan

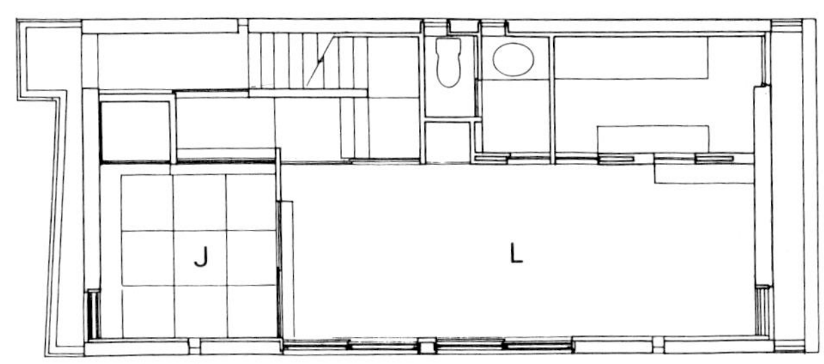

Third floor plan

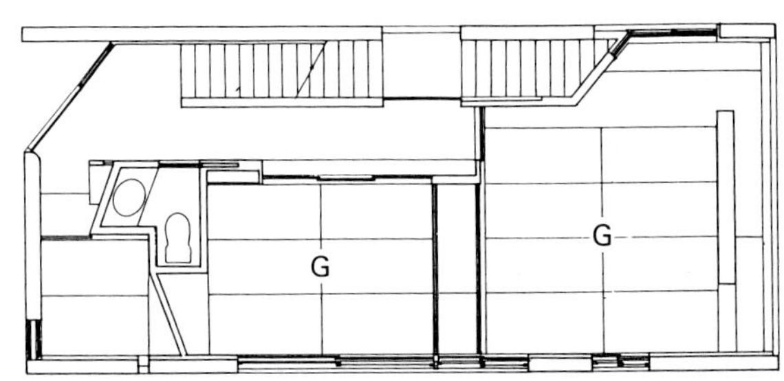

Second floor plan

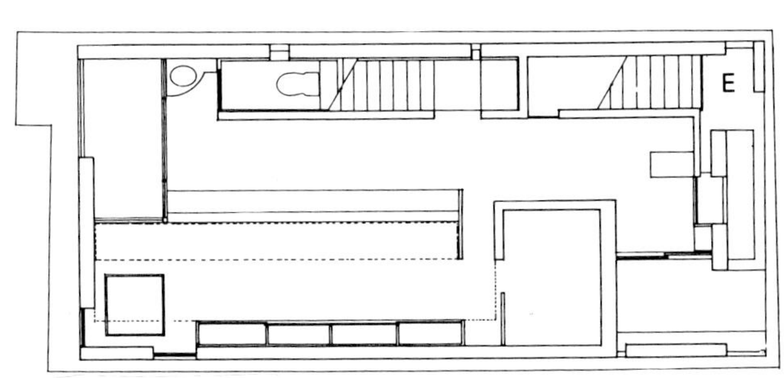

First floor plan 1/150

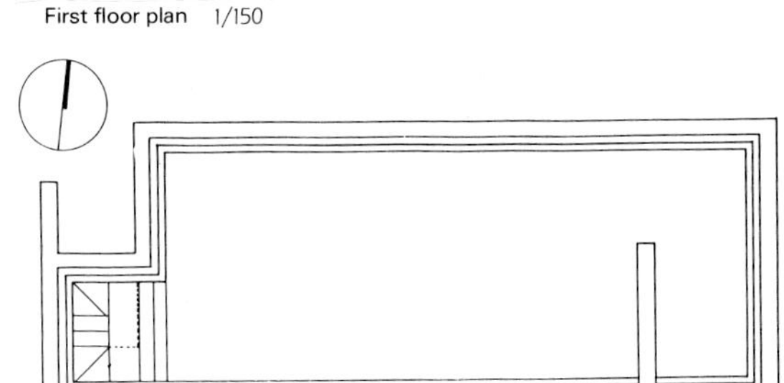

Basement floor plan

1 入口まわり。1階は料理店。左手に店への入口(開店時は板戸を内側に開けている)。右手に3階住居への入口。

2 周辺は、木造3階建ての店舗や家屋が並ぶ、下町風の懐しい町が残っている。

1 Around the entrance. The first floor is a restaurant. The restaurant entrance is on the left. (When the restaurant is open, the wooden door remains open inside). On the right, the entrance to the third floor residence.

2 In the neighbourhood, a row of three storied stores and houses. The old downtown scenery still remains.

3

5

4

6

South elevation

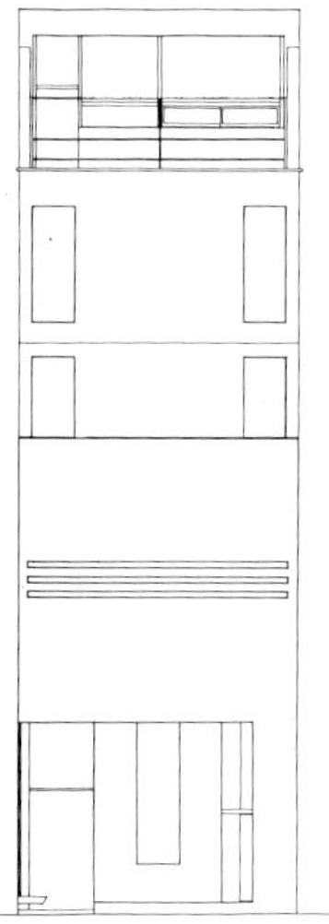

East elevation 1/250

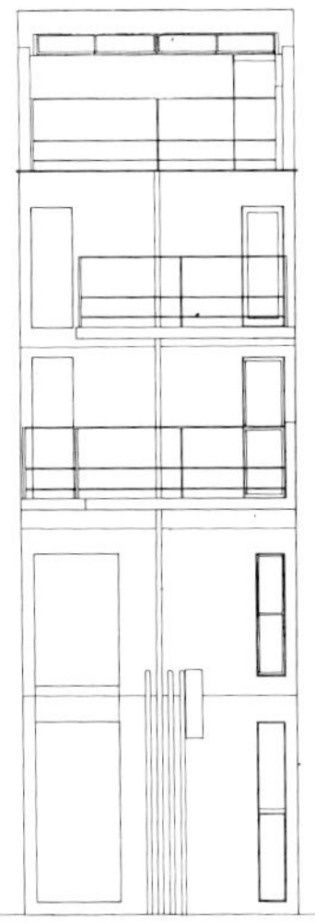

West elevation

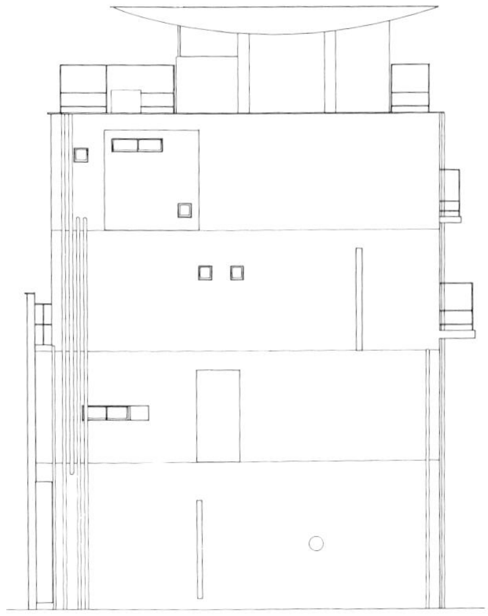

North elevation

7

8

3 周辺の建物に高さとイメージを揃えるように、2階から3階にかけて白い大理石のパネルが張り出している。
4 目立たぬように、壁と壁の間に挿入された住居への入口ドア。
5 3～5階の表情。最上階はペントハウス風である。
6 ぎっしりと建て込んだ家並みのなかで、南面は幸いにも露地に開いている。
7 8 3階。住居の玄関ドアを開けて入ったところの和室。限られたスペースのなかで、採光や細かな段差に工夫がこらされている。

3 In order to match the height and the profile, while marble panels stretch from the second to the third floors.
4 The entrance to the residence placed between walls to be inconspicuous.
5 The third to fifth floors. The top floor is in the style of a penthouse.
6 The south side opens on to the alley in the middle of a very crowded row of houses.
7 8 The third floor. The Japanese room behind the entrance door of the residence. Within the limited space, careful lighting designs takes in light and produces finely differentiated patterns.

9

10

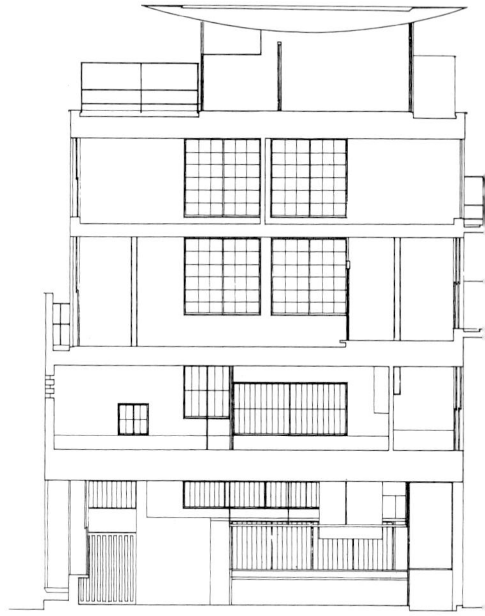

Section

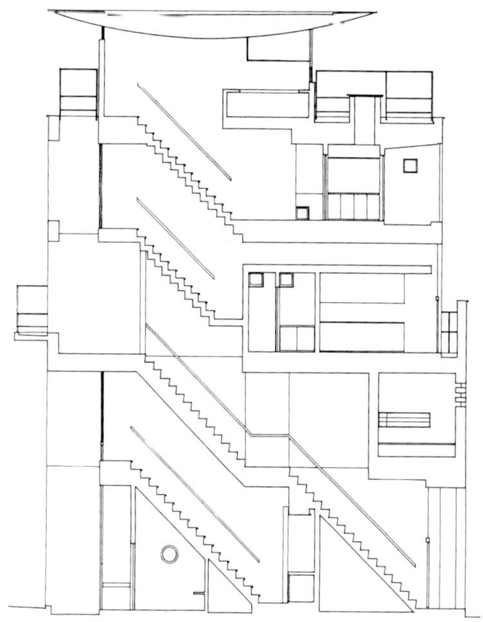

Section 1/200

9 3階。和室、居間、廊下の天井を視覚的に連続させている透明ガラスの欄間。

10 3階。和室から居間を見る。突き当りの縦長の明かり障子が、和室のそれと相対して、この住宅の端から端までの長さを見せている。床はアピトンフロ

11

ーリング貼り。
11 3階。左頁と同じ和室。和室の雰囲気を壊さぬよう、スリット状の採光窓、縦長の明り障子が使われている。同時に、打放しコンクリートや桜合板の壁が連続的に用いられる。

9 The third floor. Japanese room, living room, a transom of transparent glass to let the ceiling of the corridor continue visually.
10 The third floor. Looking at the living room from the Japanese room. Vertical light penetrates the paper sliding doors at the end contrasts those in the Japanese-style room, and shows the length of the residence from one side to the other. The floor is Apiton flooring.
11 The third floor. The same Japanese room as shown at left. In order not to break the atmosphere of the Japanese room, a slit-like window takes in light and rice-paper sliding doors are used. At the same time, bare concrete and plywood boards are used as walls.

12

13

14

15

12 4階。西端のバルコニー。ガラスブロックが階下に光を落とす。
13 4階、寝室前の廊下と階段。奥の洗面所と浴室が明るい。
14 4階。階段室と西端部を見る。
15 5階。テラスに面した書斎。曲面の天井仕上げはポリ合板。右手に、大きな箱のように置かれた装置的な和室。

12 The fourth floor. The west end balcony. Glass blocks cause the light to go through to the downstairs.
13 On the fourth floor, the corridor and the stairs in front of the bedrooms. Bright toilet and bathroom.
14 The fourth floor. Looking at the staircase and west end.
15 The fifth floor. A study facing the terrace. The finish of the vault-ceiling is resin laminate. The Japanese room placed on the right like an installation.

渡辺豊和 [WATANABE Toyokazu]

オッパイハウス 戸口邸 [The Breast House]

■渡辺豊和の建築の原形質を知ろうとするならば、「餓鬼の舎」と名づけられた自邸を訪ねるのが一番だろう。正確にはアネックスであり、ほぼ正方形プランの木造の母屋(これも渡辺が以前設計した)の2面を、幅2m、一辺の長さ10mあまり、高さ3層・L字形プランで囲う、打放しコンクリートの奇妙な建物である。間仕切ひとつない、ほとんど通路だけでできているような内部は現代の一般的住宅からもっとも遠い位置にあり、同時に決して実現されぬはずだった幻想的空間であるといえる。つまり、このアネックスが実際に建てられたこと自体が驚くべきことなのだが、そのまま現代住宅に対する根本的な批判的存在となっているのである。彼の方法論が直截に見える住宅群、例えば1と½のドームの下に置かれた和風の茶室、あるいは和風の家の中に隠されたギリシャ神殿といった、重層的構成による幻想建築は、ごく早い時期から、現代建築を過渡的な一様式と見ていたスタンスからつくり出されている。
打放しコンクリートは、自分にとっては石の表現ではなく、土を意味している、と渡辺は語ったことがある。ロマンティックな廃墟嗜好などではない、より地底的な世界へのボーリング。最近作の「神殿住居」では、地底的なものと、重層的な手法と、それら総体を宇宙模型として見せる、トータルなイメージにまで達している世界の実現がある。

■版画家である施主の、地階ホールの陶画に見られるような、どことなくユーモラスな画風を反映して、この住宅は設計されたという。建物を見ると、グロテスクで異様な味つけが印象的だが、渡辺にとっての、現実的な解答の仕方だろう。健康で微笑ましいユーモアをそのまま建築にしたとしたら、薄気味の悪いものになることは間違いない。
道路に溢れ出た、フェリーニ的な巨大性をもった女性の胸部。その上を覆うより大きな半円の屋根は、内部においてはさらに完結を求めて家の奥に広がり、孔部のある内膜のようなドームの一部を現す。こうした大きなディメンションのエネルギーは家の頂上まで一向に衰えず、この孔のあいた浅いドームを扁平にした形の、薔薇窓のような開口部をもった真円が、個室の断面を決めている。この真円は、同じ大きさのはずみ車状のフレームとなって、屋根の反対側の端部に聳え立つ。そのシルエットは、古いミシンのようだ。打放しコンクリートは、ここでは人体のような感触を帯びる。巨大な人間の器官のなか、あるいは人体が柔らかい機械へと化していく、その迷宮の内部である。それは荒涼たる夢の場所ではない。乳房から浅いドーム、「薔薇窓」、そしてその骨格を残したはずみ車へと、同一のモチーフが丹念に反復されて、思いがけない住宅内部の広がりをつくり出す。幾何学精神が、人間の身体と建築の空間との謎めいた同質性に向かっているのである。

■The best way to know the architecture by Mr. Toyokazu Watanabe is to visit his house, named the "Brat House". It is an annex to the wooden main wing of an almost square plan (previously designed by Mr. Watanabe). Two sides of the wing are enclosed by the L-shaped annex which 2 meters wide, 10 meters long on a side and three-stories high. The house is a queer building of ferroconcrete. There is not one partition wall in the house; the entire house looks like a makeup of corridors. The inside of the house is far from current general residential buildings, and the nearest to a phantom space which could ever be realized. Not only it is a surprise in itself that this annex was ever built, but the existence of the annex is a critique on the principles of modern houses.
Other houses from which his methodology is more plainly visible include one in which a Japanese tea ceremony room is placed under the dome of the first floor, or one in which a Grecian sanctuary is hidden in a Japanese style room. The phantom architecture of this multistoried structure has been produced from his stance at an early time when modern architecture was considered a transient style.
Mr. Watanabe once said that the ferroconcrete surface did not express stones, but it meant soil to him, — not a romantic fondness for ruins, but something from the deep earth. In his recent work "Divine House", there is a world of total images in which things of the ground and multistory techniques are wholly expressed in a space model.

■It is said that this house was designed to reflect something of humorous printing style of the owner, who is a woodblock artist. His porcelain pictures are displayed in the basement hall. From the building, you may get an impression of something grotesque and in peculiar taste, but this may be Mr. Watanabe's style of realism. If healthy and smile-provoking humors are converted into a building as they are, it will surely be a strange building.
Huge women's breasts overflow the road. A larger semi-circular roof covering these breasts expands deeper into the house for more completeness and expresses a part of the dome which is like an lining membrane with a foramen. This vast energy does not decrease a bit moving up to the top of the house, and perfect circles having openings like rose windows define the cross sections of individual rooms. These circles become wheel-shaped frames of the same size and rise high above the edge of the roof. The silhouette is like an old-fashioned sewing machine.
The ferroconcrete wears the touch of the human body. The inside organs of a giant or a mechanization of a human body — the interior of a labyrinth. It is not a place of a desolate dreams. The motifs from the breast, to the low dome, to the "rose window", to the framed wheel are elaborately repeated and create a vastness in the dwelling. The geometrical spirit faces the mysterious homogeneity of the human body in the space of a building.

1 南西側から見る。

1 A view from the southwest.

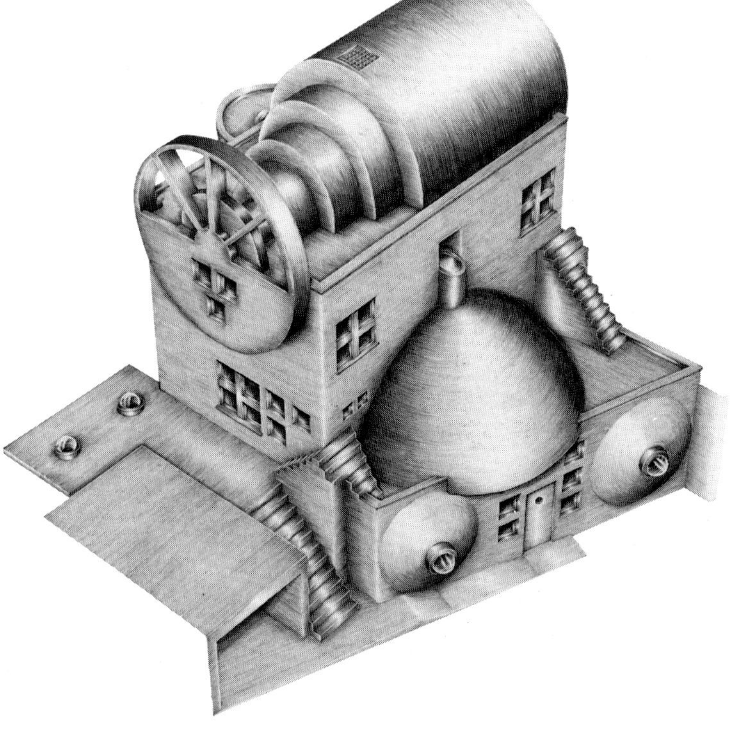

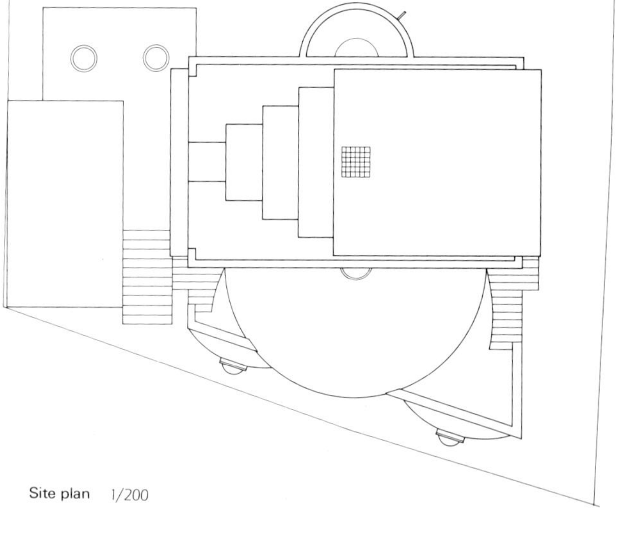

Site plan 1/200

1

2

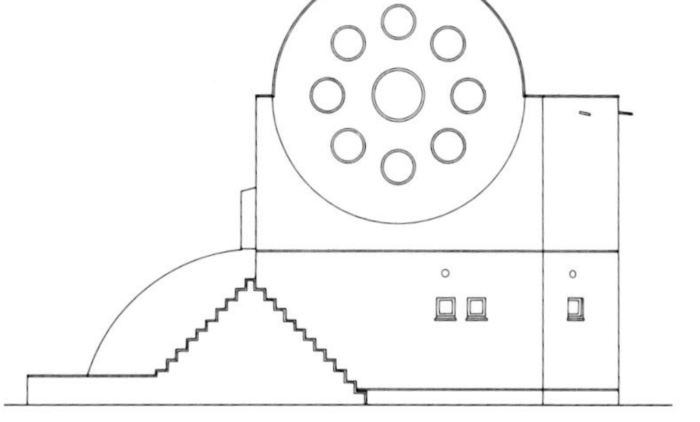

South elevation

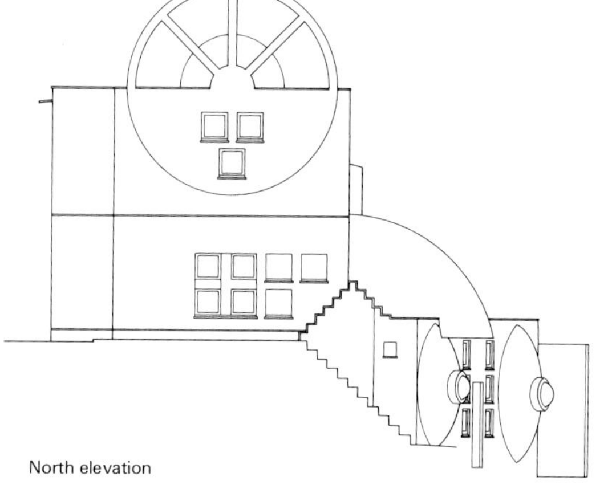

North elevation

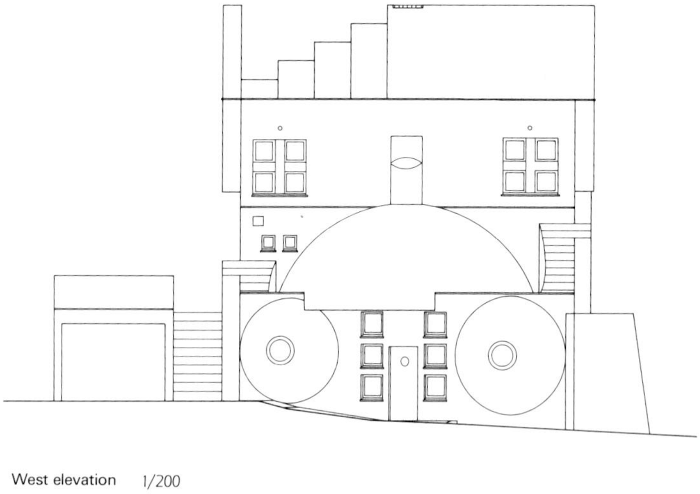

West elevation　1/200

2「オッパイ」のクローズアップ。
3北面のエレヴェーション。玄関は地下１階になっている。

2A close-up view of the "Breast".
3The elevation on the north side. The foyer is in the first basement.

3

4

5

6

7

4 地下1階。ホール。右手に玄関ドア。陶壁は、施主の戸口勉氏による。
5 左の「オッパイ」を内側から見る。
6 7 ホールは中央の階段を狭んで左右にくびれているが、天井は一体感を強調している。

4 The hall in the first basement. To the right, the front door. The ceramic wall is made by the owner, Mr.Tsutomu Toguchi.
5 The left "Breast" viewed from the inner side.
6 7 The hall is constricted on both sides facing the center stairway, but the monolithic construction is emphasized on the ceiling.

8

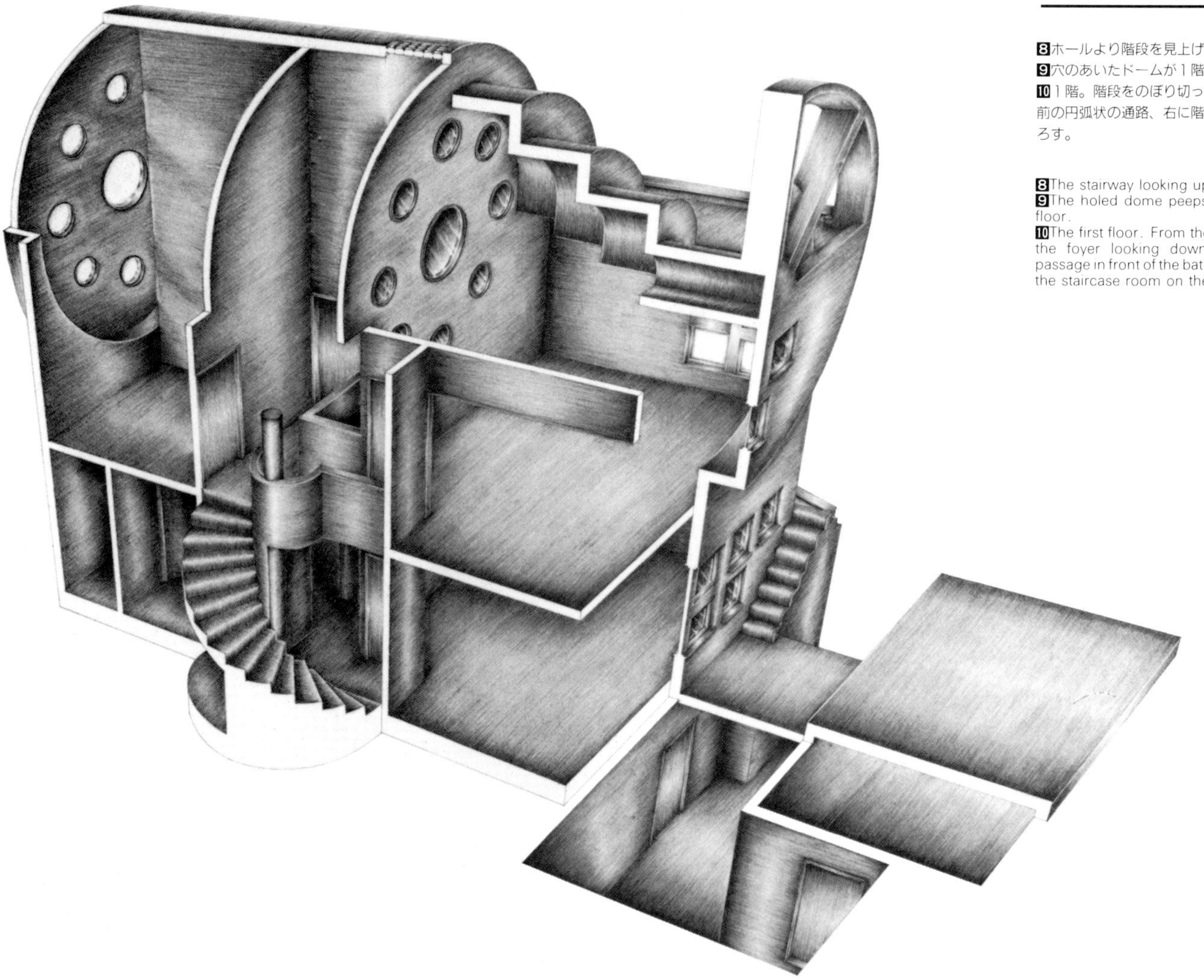

8ホールより階段を見上げる。
9穴のあいたドームが1階フロアに顔を出している。
10 1階。階段をのぼり切ったところから、左に浴室前の円弧状の通路、右に階段室を通して玄関を見下ろす。

8The stairway looking up from the hall.
9The holed dome peeps its figure in the first floor.
10The first floor. From the end of the stairway, the foyer looking down through a circular passage in front of the bathroom on the left, and the staircase room on the right.

9

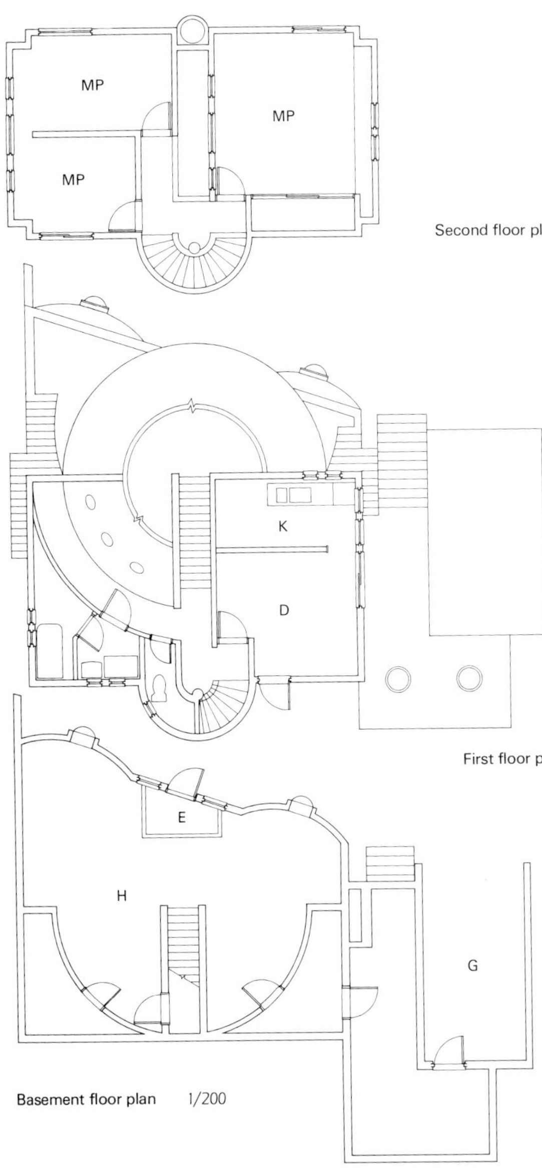

Second floor plan

First floor plan

Basement floor plan 1/200

10

11

12

13

11 12 階段室の吹き抜けに面した「薔薇窓」。内側に主寝室。

13 1階と2階を結ぶ階段室。

14 外に開いた、もうひとつの「薔薇窓」。2つに区切られて子供室になっている。

11 12 The "Rose Window" facing the well in the staircase room. Inside is the master bedroom.

13 The staircase room connecting the first and second floors.

14 Another "Rose Window" opened outward. Inside are two partitioned children's rooms.

14

神殿住居 地球庵[Divine House "Chikyuan"]

■この住宅は、石井修の住宅群が緑と石のなかに隠されている目神山に、今度はこれ以上ないほど露出して聳えている。敷地は道路から7mを越える高さの崖である。アクセス階段を蜿々とのぼっていった上に、打放しコンクリートの壁と2つのドームが巨岩のように積み上げられている。ここには、深い森がある一方、確かにこのような山頂の巨岩群があちこちに見られ、2人の建築家は、それぞれ目神山の自然から自分のつくるべき建築を探し当てたに違いない。
渡辺自身の説明によると、この家を設計していた記憶が全く欠けており、図面を描いた覚えもなく、当然残っているべきエスキースも1枚も残存しないという。この家には施主の意志も建築家の意志も反映されていず、地霊の誘うまま、神の差配によってのごとく、実現していたというのである。ロマン派やシュルレアリズムの画家や音楽家や詩人たちが、夢の裡になした作品と共通する、ひたすら精神的な要因を、渡辺はこの住宅の成立に託しているように思える。
その真疑を問う以上に、この夢の家は、現代の住宅を成立させる要因の一切を廃棄してはじめて現れてくるようなポレミックな存在形式を獲得している。それは球形の茶室「地球庵」はいうに及ばず、すべての部屋々々とその結合部分が雄弁に語っている。理想化された世界ではなく、得体がしれないものの確かさ。
渡辺はその存在に対して、「神殿住居」と名づけずにはおれなかった。

1

■ This house rises above the peak of Mekamiyama, exposing its entirety, where houses designed by Mr. Osamu Ishii are hidden among the greens and stones. The site is on a cliff seven meters above thc road. At the top of a winding access stairway, ferroconcrete walls and two domes are piled up like huge rocks. Here and there at Mekamiyama, there are deep forest as well as groups of huge rocks. The two architects must have each found his own architecture from the nature of Mekamiyama.
According to Mr. Watanabe's explanation, he does not remember at all designing this house, nor does he recall drawing the plans. He says that not one piece of design work exists. He further said that this house did not reflect the will of a owner nor the will of an architect, but was materialized by the soul of the soil and by the direction of gods. There must be something common here with the works of romantic and surrealistic painters, musicians and poets which finally materialized after days of long-dreaming.
Aside from examining its truth or falsehood, this house of dreams has a polemic existence which becomes clear only when all the rules of modern housing are discarded. Not only the spherical tea-ceremony room "Chikyuan" but also the other rooms and their adjoining portions of eloquently speak of this fact. The existence of the "Chikyu-an" is in a world of undefinables, not an idealized world.
Mr. Watanabe could not but name it "Divine House".

Site plan 1/300

2

❶道路から40段以上の階段をのぼって、やっと玄関にたどり着く。
❷北西からの全景。

❶After a climb of more than 40 stairs from the road, we reach the foyer of Chikyuan.
❷Overall view from the northwest.

3

3「序の間」(玄関わきのホール・右手)と階段室。のぼり切ったところに「地球庵」がある。
4東側外観。右下に玄関。地下1階に当たる。
5「序の間」。左手に階段室。右手、黄色いドアの向うに、「遊の間」(子供室)。
6「遊の間」のドームを見上げる。

3"Prelude Room" (to the right of the hall beside the foyer) and staircase. At the top step, there is a "Chikyuan"
4An exterior view of the east side. There is a foyer at the bottom right-hand corner, located in the first basement.
5The "Prelude Room". To the left, the staircase room, and the "Play Room" (the nursery) beyond the yellow door on the right.
6The dome of the "Play Room" looking up.

4

5

6

7

8

9

7「遊の間」のドームを見上げる。写真上方は「序の間」。
8「序の間」から階段室の天井を見上げる。右は「序の間」の天井。
9地下1階。「序の間」。階段の足元のコーナー。
10地下1階。「地象の間」から「遊の間」を見る。2つの「間」は、床は連続しているが、「地象の間」が天井の高さを1階分よりさらに低く抑えているのに比べて、「遊の間」は2階分吹き抜けの上にさらにドームが乗っている高さ。
11 1階。「序の間二」。2階へ真直ぐ上がっていく階段室の踊り場的なホール。コンクリートの間柱を通して、「序の間」の吹き抜け上部が見える。
12 1階。「心象の間」(居間)から2段分降りた「宴の間」(台所・食堂)を見る。「地球庵」(茶室)の下半分が天井から下がっている。

7The dome of the "Play Room" looking up. The top of the photograph is "Prelude Room".
8The ceiling in the staircase looking up from the "Prelude Room". To the right, the ceiling of the "Prelude Room".
9The first basement floor. The "Prelude Room", a corner at the foot of the stairway.
10First basement floor, the "Play Room" as viewed from the "Material Image Room". The two rooms are on the same level, but the ceiling of the "Material Image Room" is lower than a one-story ceiling height, while the ceiling height of the "Play Room" is a two-story well with a dome above.
11The first floor: The "Prelude Room 2", the landing of the staircase leading straight to the second floor. The top of the well in the "Prelude Room" as seen through a concrete principal pillar.
12The first floor. The "Feast Room" (kitchen and dining room), viewed from two steps below the "Mental Image Room". The lower half of the "Chikyuan" (tea-ceremony room) is seen below the ceiling.

10

11

12

13

13 1階。「宴の間」。「地球庵」の下半分が食卓の真上に迫る。この食卓の位置は、施主夫人が決めたという。

14 16「地球庵」の上部にリング状の回廊がある(「天境の間」)。さらにその真上にドームが重なっている。

15「地球庵」をめぐる回廊。

13 The first floor. The "Feast Room". The lower half of the "Chikyuan" is just above the dining table. The position of the table was suggested by the owner's wife.

14 16 The ring cloister (the "Heaven| Boundary Room") at the top of the "Chikyuan". The cloister is domed.

15 The cloister surrounding the "Chikyuan.

14

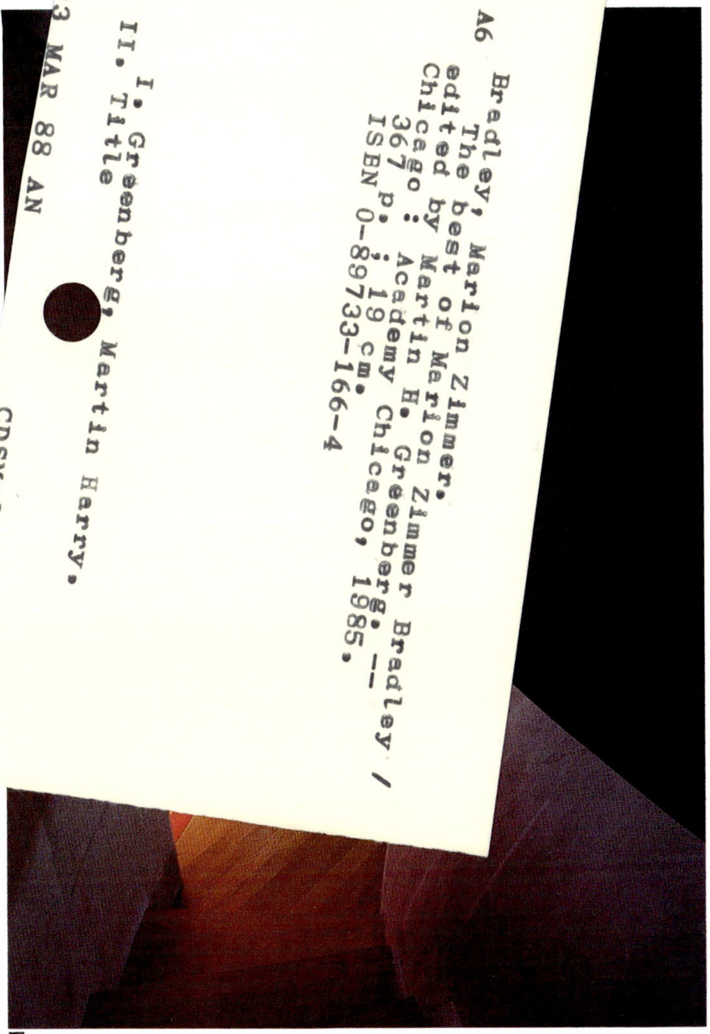

15

16

17

18

17「天境の間」より「地球庵」を見下ろす。この回廊はそのまま書斎となって北にウィングを伸ばしている。
18中3階。書斎。
19屋上。円弧状にカットされた壁と屋根がドームを抱く。
20「神殿住居」の名に相応しい謎を帯びた表情。

17"Chikyuan" looking down from the "Heaven Boundary Room". This cloister becomes part of the study and extends the wing to the north.
18The semi-third floor. Study.
19Rooftop. The circularly cut wall and the roof outlines a dome.
20A mysterious expression worthy of its name "Divine House".

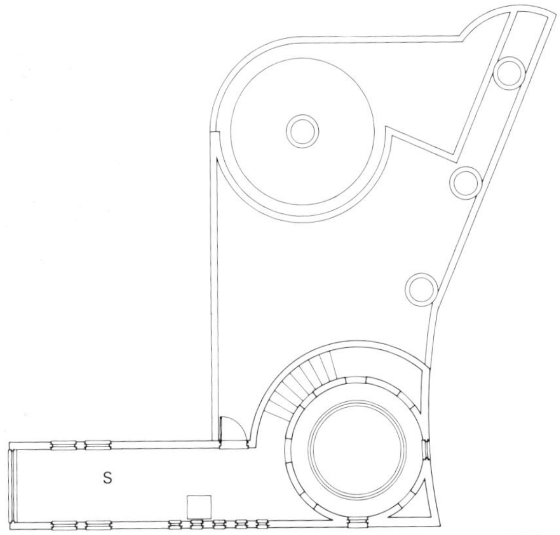

Mezzanine between the second and third floors

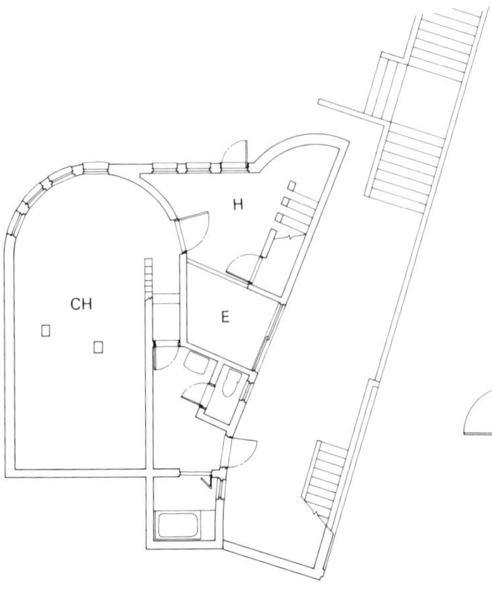

Basement floor plan

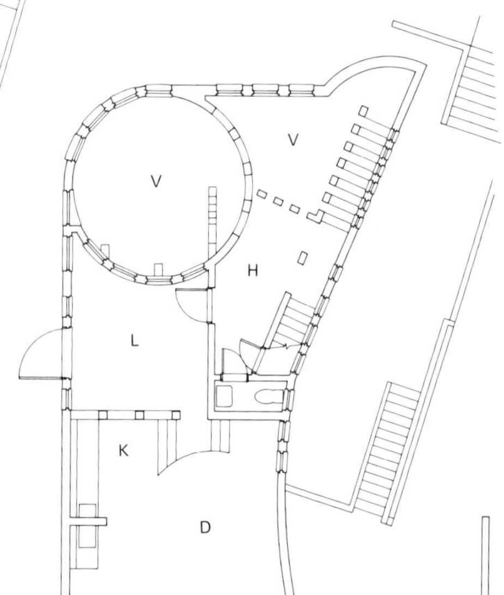

1/200
First floor plan

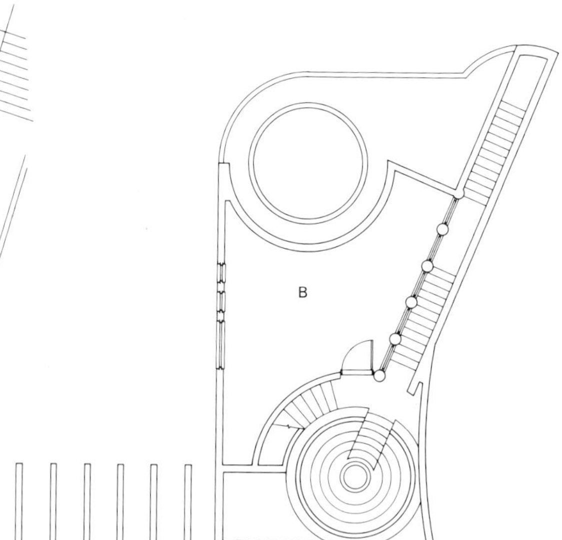

Second floor plan

19

20

山本理顕 [YAMAMOTO Riken]

GAZEBO

1

■戦後から現在にかけて、日本の住宅は欧米的なL＋nBのプランニングを定着させてきた。このことに対する批判を、実際の住宅設計を通じてもっとも鮮やかに示してきたのが、山本理顕である。
彼は日本の一般的家庭における生活の実態調査を行い、それを構造化している。さらに地中海やアフリカの集落調査を通して、理論的な補強を行ったうえで、ここ10年の設計活動に入った。
彼の住宅に一貫しているのは、プランニングが要求する建築の形態という問題をつねに迂回してきたこと、端的にいえば、部屋の間取りと外観とが無関係であるような状態を保ちつづけてきたのである。だから1つひとつの住宅はすべて違うスタイルを見せ、切妻の和風であったり、地中海の民家風であったり、体育館のようだったりする。
このことで、彼が建築家として自分のスタイルを確立することに無関心であり、あるいは移り気な作家であると受けとられることさえある。しかし、彼は、この領域における理論と表現との関係が、とんでもない広がりをもつ事実を示しているにすぎない。佐藤邸においては、部屋の配置をそのまま家の外形にした結果、その打放しコンクリートのマッスは、プエブロインディアンの土の住居のようになった。そのヴァナキュラーな様相にも、山本は安定した住宅の形態を見ない。彼は、全く別の位相に隠されている「建築」――つまりプランニングと形態との本来的な関係を探り出そうとしている。

■横浜の市街地、幹線道路に面して建つ雑居ビルである。1階は2軒の店舗と駐車場、2階は貸事務所、3階は貸アパート4戸。それにオーナーの住居への玄関、書斎、個室(寝室)。このゾーンをゲートとして、屋外の鉄骨階段をあがった所、4階全体がオーナーの家族の住居である。しかし、ここには1軒の家があるというより、家族室、2つの個室、浴室等の部屋々々が屋上に並べられているといったほうがいい。部屋々々をまとめる動線もシェルターも外壁もない。まとめているものがあるとすれば「入口(ゲート)」のゾーンだけである。
これらの部屋は、本質的にバラバラに散在する。城塞のように堅固な壁で囲われていても、家族の1人ひとりが独立して社会と関わっている以上、個々の部屋は外に対しても開いているという考えである。つまり、山本の住宅は、家に対する幻想を捨てたところから成立している。
この家の屋上には、四囲の都市の光景が、おそるべきエネルギーで流れ込んでいる。部屋は都市のなかにバラバラに浮きながら、それでもそこに家族の住まいをまとめる手法があるという、幻想を捨てた果ての確信に満ちた構成。
打放しコンクリートを主体としながら、ステンレス織のメッシュとテントで、壁と屋根とを軽やかに囲って、内部構成の強さを逆説的に示している。

■From the end of World War II to the present, the L+nB planning in Europe and America has been planted in Japanese housing. Through his designing, Mr. Riken Yamamoto has shown his clear criticism of this planning in Japan.
He conducts fact-finding surveys on living conditions in general Japanese households, and expresses them in his constructions. He entered his 10-year long designing activities after he had strengthened his theories through investigations of communities in the Mediterranean and Africa.
His consistent idea on housing is that the room arrangement in a house is unrelated to the external appearance of the house. Therefore, the houses he designs all show different styles, some have Japanese gable styles, some have the style of private houses in the Mediterranean, and some look like gymnasiums.
Because of this, it is sometimes thought that he has no interest in establishing his own style as an architect or that he is whimsical. But, he is only showing the fact that the relationship of theory and expression in the construction field is unexpectedly expandable. As a result, the room arrangement of the Sato House was made exactly the same as the external shape of the house, its ferroconcrete mass became like a mud house of a Pueblo Indian. Mr. Yamamoto does not see a style of stable house in its vernacular appearance. He is trying to search for "architecture" hidden in a completely different phase.

■A building housing many independent business institutions stands facing the main road in an urban district of Yokohama. Two stores and a parking garage on the first floor, rented office rooms on the second floor, and four apartments for rent on the third floor. Further, there are the foyer to the owner's house, his study and private room (bedroom) on the third floor. Making this zone a gate and climbing up the outdoor steel-frame stairway, the entire fourth floor is used as a house by the owner's family.
It may be better to say that rooms, a family room, two private rooms and a bathroom, are lined up on the rooftop, rather than to say that one house exists in the building. There is not a path of flow to combine rooms, nor a shelter, nor an enclosure wall. There exists only the "gate" zone to combine the rooms.
These rooms exist essentially apart from each other. Even if independent rooms are enclosed by a rigid wall such as in a fortress, individual rooms open outward as long as each member of the family is independent and concerns himself in society. In short, Mr. Yamamoto's housing starts from when a fantasy of houses is abandoned.
Above the rooftop of this house, this spectacle is flowing in with frightening energy. A fantasy created after the abandonment of a fantasy rooms sporadically float in the town and still a family dwelling is maintained.
Although the building is covered mainly with ferrconcrete, its walls and roof are enclosed with stainless steel mesh and a tent, paradoxically exhibiting the strength of the internal structure.

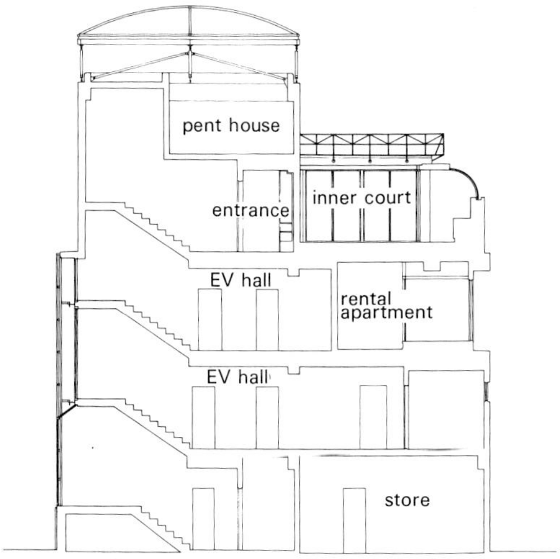

Section

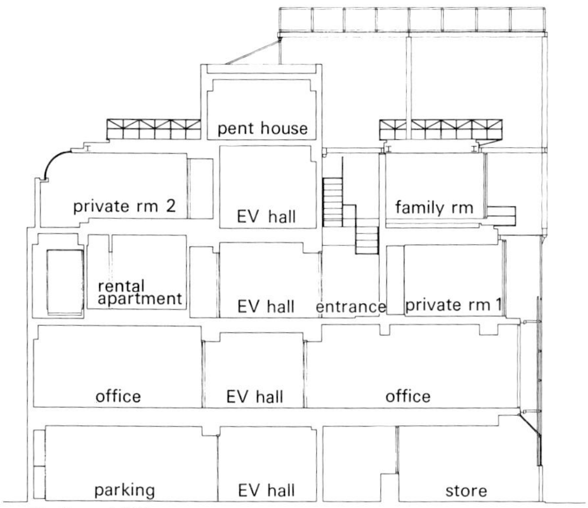

Section 1/500

2

3

1南東側全景。道路に面した姿。
2北西側から見る。西側に路地がある。
3南西側から見上げる。2階までと3階の一部のファサードはステンレス・メッシュで蔽われている。その上部に住居のバルコニーの積層が出現する。

1Overall view of the southeast side, facing the street.
2Viewed from the northeast. An alley on the west.
3Viewed up from the southwest. The facade, the second floor and part of the third floor is covered with a stainless steel mesh. Above this, rows of balconies attached to each house appear.

4

5

6

456 3階にはアパート3戸とオーナーの住居の一部がある。エレベーター前の鉄のドアを開けると、突然目の前に外階段が現れる。住居の玄関ホールでもある。階段の足元に、管理室のような書斎。
7 4階。住居の中庭と、それに面した家族室や浴室。

456 The third floor has three apartments and part of the owner's dwelling. When an iron door in front of the elevator is opened, scene suddely appears before your eyes. This is the entrance hall of the owner's house. At the foot of the stairs, there is a study which looks like an office.
7 The fourth floor. The inner garden of the owner's house, and the family room and bathroom facing this garden.

7

8

9

10

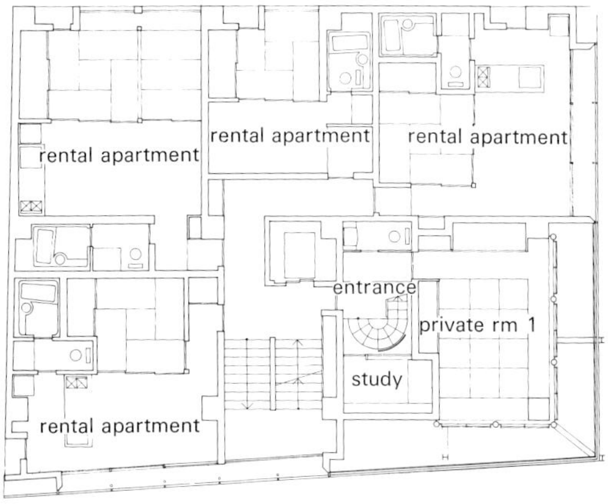

Third floor plan

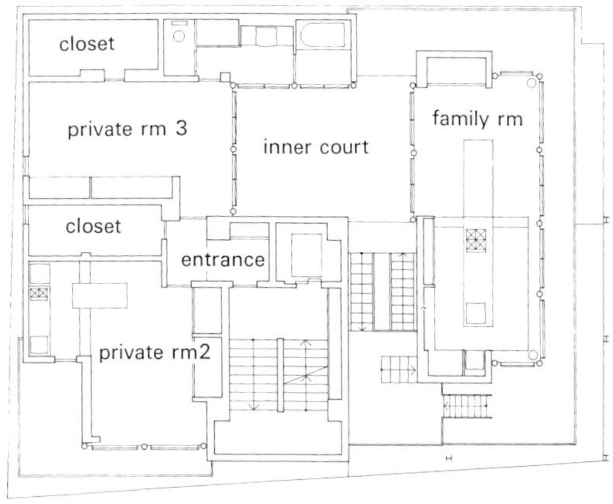

Fourth floor plan

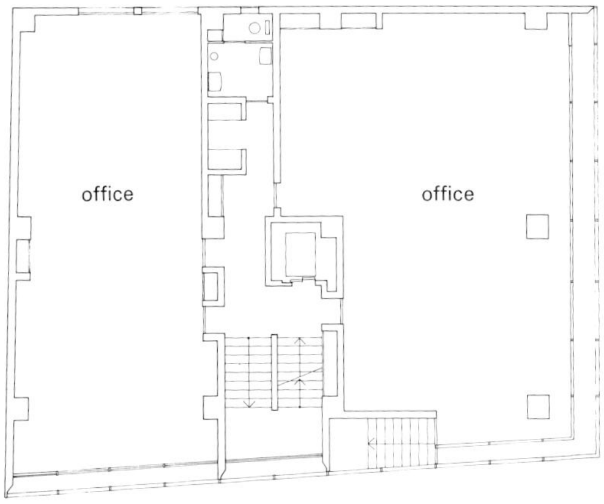

Second floor plan

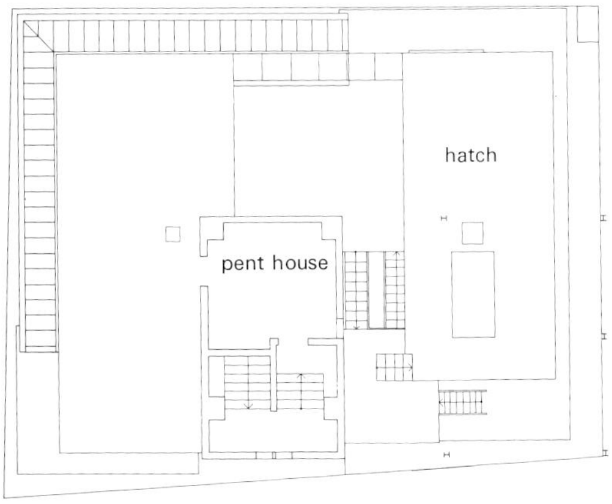

First floor plan 1/500

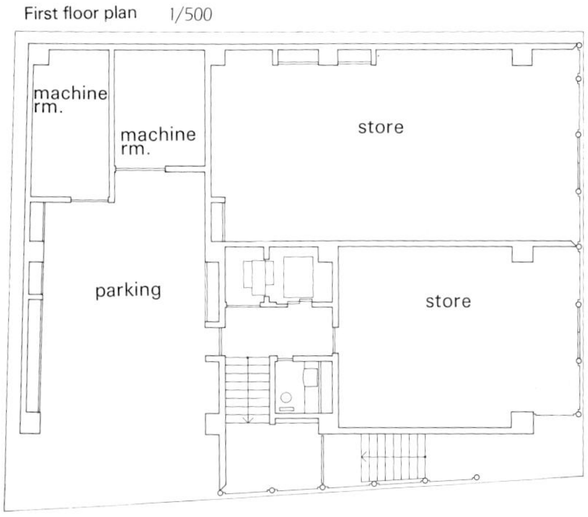

Basement floor plan

11

8 4階。家族室から中庭を見る。中庭は檜の簀の子縁。

9 中庭に面した家族室。右手に、3階からの外階段。

10 中庭に面した浴室と便所。屋外にありながら、「家の内部」の印象を強く与えるエレメント。

11 階段をさらに上に上がって行くと、家族室の屋上に出られる。

8 The fourth floor. The inner garden which has a hurdle verandah of Japanese cypress, viewed from the family room.

9 Family room facing the inner garden. To the right, the outside steps rising up from the third floor.

10 Bahtroom and washroom facing the inner garden. The garden provides a strong impression of being inside the house though it is located outside.

11 Further up the stairs leads to the rooftop of the family room.

12

13

14

15

12家族室の屋上から中庭を見下ろす。エレベーターと階段のコンクリート・シャフトを中心に、鉄骨造の部屋が集まっている。
13コンクリート・シャフト。機械室の通風孔が中央に見える。
14コンクリートの壁、鉄骨とガラス、中庭の簀の子、そして天蓋のテント。
15積層したバルコニーの表情は、強く、アノニマスで、かつ幻想的だ。

12The inner garden viewed down from the rooftop of the family room. Centering the concrete shaft of elevatior and stairs, steel-frame houses congregate.
13The concrete shaft and a vent hole for the machine room are seen in the center.
14Concrete wall, steel-frame and glass, grillage in the inner garden and canopy.
15The expression of layered balconies is strong.

16

17

18

16 南側道路より、住居部分を見上げる。3階和室とその上部の家族室。
17 西側ファサードを見上げる。ステンレス・メッシュとガラス面。その上部に突き出たコンクリートの階段室と住居のバルコニー。
18 南側道路に面したファサード。天蓋を支える鉄骨の高さはほぼ10m。

16 The dwelling looking up from the south street. Japanese style room on the third floor and family room on the upper floor.
17 The facade on the west looking up from the lower level. Stainless steel mesh and glass, above which a concrete staircase and balcony for the dwelling protrude.
18 The facade facing the south street. The height of the steel frame supporting the canopy is approximately 10 meters.

■「ガゼボ」のすぐ近く、やはり車の往来の激しい街路に面して建つ、これも雑居ビルである。1階に店舗、駐車場、2階に貸事務所、3階に貸アパート4戸、4、5階にオーナーの住居という構成も同じだし、道路に面したファサードが南面であるためか、建物の表情がはっきりと道路に開いている点も同じ。しかし、こちらはよりいっそうその表情がダイナミックになり、「ガゼボ」が街のなかに目立たずあるのに比べて、やや挑戦的でさえある。4階から上の住居部分は、もはやペントハウス的なプロポーションを越えて、建物の上半分を占める支配的な顔となり、「ロトンダ」の名に相応しい、古典主義的な鉄のオーダーを誇示し、一方、地中から迫り上がってきたかのような打放しコンクリートのマッスと拮抗している。
打放しコンクリートの、それこそプエブロインディアンの、空に剥き出しになった古代的な住まいに、近代工業化時代の産物のような鉄の多様な表情が重ねられ、さらにその上部に巨大なテントが重なる。既視感を誘い出し、同時にSF的でもある。ヴァナキュラーな様相が、建築の手を借りずに「建築」として凝縮する。それを可能にする方法のひとつとして、現在の山本は、単なるシェルターを超えた「ルーフ」に注目している。

1

■This is also a mixed building of houses and offices built near the "GAZEBO" facing a busy road. Stores and a parking garage on the first floor, rented offices on the second floor, four apartment houses for rent on the third floor and the owner's house on the fourth and fifth floors. Its makeup is same as the "GAZEBO". Another similarity is that the appearance of the building clearly opens toward the road probably because the facade of the building facing the street are on the south. The appearance is more dynamic and provocative in comparison with the "GAZEBO" standing in the shade of the town. The residential portion above the fourth floor occupies the upper half of the building, exceeding a penthouse in proportions, and exhibits a dominant face, worthy of the name "Rotunda". It shows off an iron classicism, while combining with a ferroconcrete mass as if it was raised up from underground.

Over the ferroconcrete, this is an ancient dwelling looking toward the sky. Various expressions in iron, a product of the modern industrialized times, overlapped and a huge tent covers the top. The building seduces visual sensations while being scientifically fictional.

The vernacular is condensed into "architecture" without the help of architecture. As one methods of making this possible, Mr. Yamamoto is now directing his attention to unfunctional "roofs".

2

1 南側道路からの全景。
2 4、5階の住居部分は、天蓋を支える鉄の円柱によって明示されている。

1 Overall view from the south street.
2 The dwelling on the fourth and fifth floors is clarified by an iron round pillar supporting the canopy.

3

3天蓋と列柱によって包まれ、同時に外にそっくり現れている住居の形。
45一方、建物の側面では、半透明の金属のスキン(ステンレス織網)が印象的だ。

3The dwelling is covered with canopy and peristyle. Its outer shape is clearly visible.
45Semitransparent metal skin (stainless steel mesh) on one side of the building produces an impressive scene.

4

5

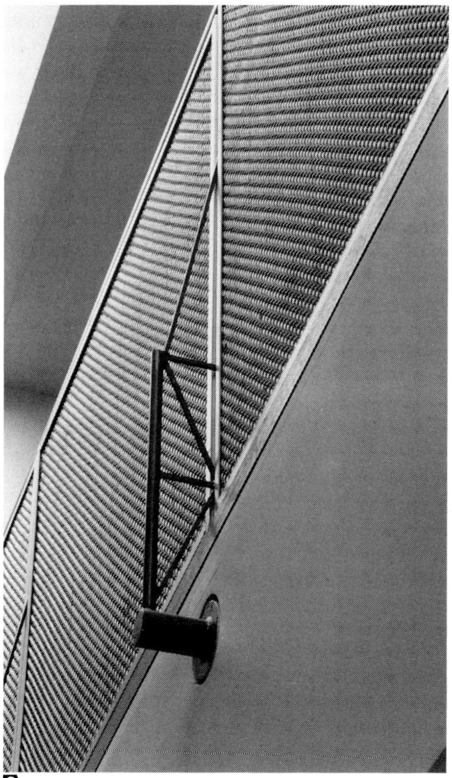

6

7

8

6→11 鉄と鉄、鉄とコンクリートとの取り合わせ。

6→11 Combinations of iron and iron, and iron and concrete.

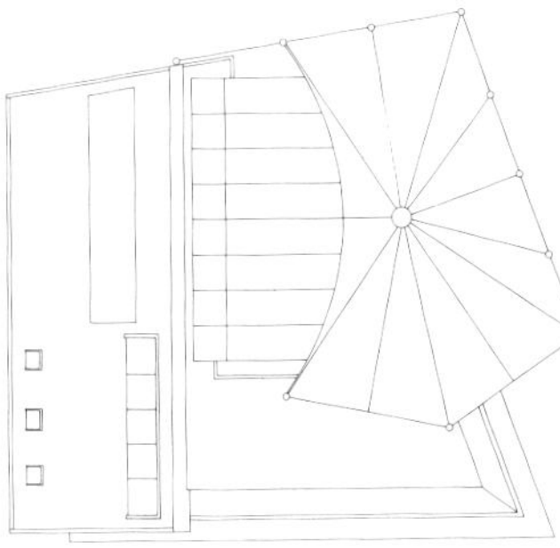

Rooftop plan

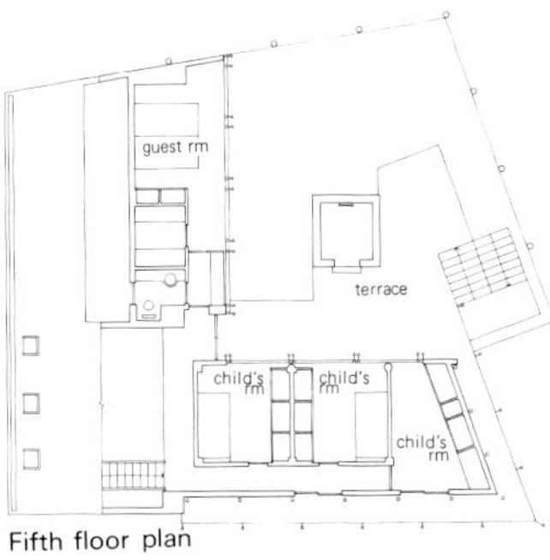

Fifth floor plan

Fourth floor plan

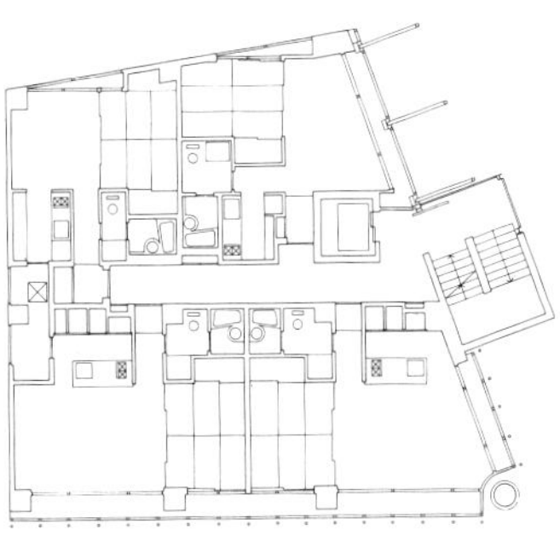

Third floor plan

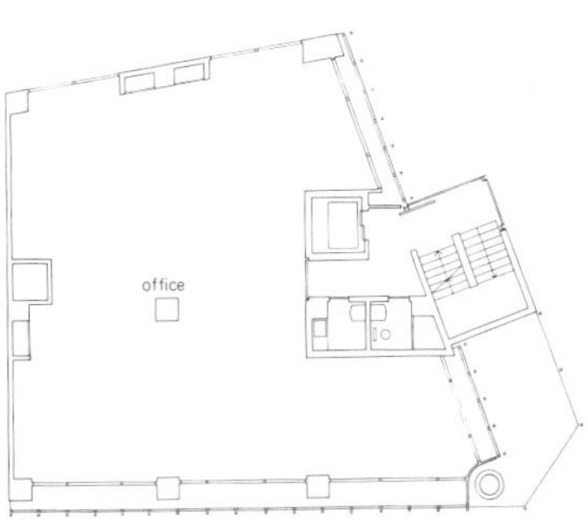

Second floor plan

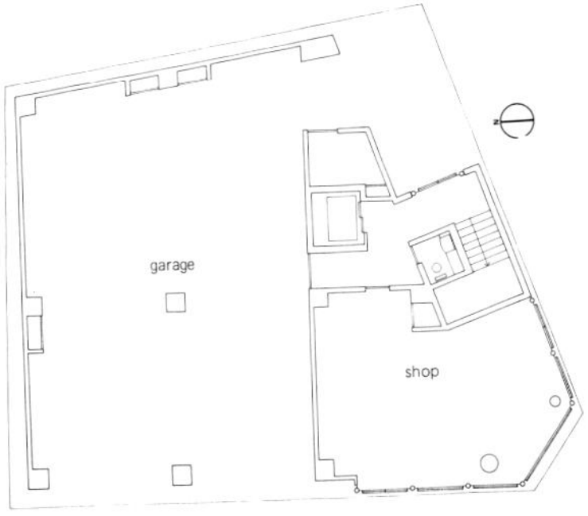

First floor plan 1/800

9

10

11

12

13

14

15

12 4階。ファミリー・ルーム。
13 ファミリー・ルームから、5階個室階の廊下を見上げる。廊下に沿ってスカイライトが開かれている。
14 ファミリー・ルームとその奥に続く和室。天井のパイプは、照明器具のサポート。
15 十二帖の和室の中央は襖だけで仕切られている。ゲスト・ルームとして活用するために、水まわりも完備している。

12 The fourth floor. Family room.
13 Corridor between private rooms on the fifth floor viewed from the family room. Along the corridor, skylights open.
14 Family room and Japanese style room in the back. The pipe on the ceiling supports lighting equipment.
15 The 12-mat Japanese style room is partitioned only at center with "fusuma" sliding doors. Because the room is also used as a guest room, a water-supply is provided.

16

17

18

16 17 5階。子供室前のテラス。
18 4階のテラスより、5階子供室前のテラスを見上げる。
19 エレベーター・シャフトの上には、グレーチングの展望台が乗せられている。さらにその上部に、巨大なテフロン膜のテント。

16 17 The fifth floor. Terrace in front of the children's room.
18 Terrace in front of the children's room on the fifth floor, looking up from the terrace on the fourth floor.
19 Above the elevator shaft, a grating view point, above which there is a huge tent coated with Teflon.

19

吉田保夫 [YOSHIDA Yasuo]

田中邸 [Tanaka House]

■吉田保夫の、打放しコンクリートは砂のヴィジョンをもつ。それも、流れる砂、あるいは限りなく崩れていく砂。
彼は実際、砂を使う建築家である。喫茶店やバーの床全面を砂で盛り、客の靴を白くまぶしてしまうし、ある演劇の舞台美術を手がけたときは、天井から滝のように砂を落とし続ける手法を用いたりしている。そして、彼の建築においてはコンクリートも鉄も紙も布も、すべてが崩れ、錆び、裂けていくような表情が主調となる。あるいは部屋と部屋との結びつきも不安な様相を呈し、隣り合った部屋の片方が突然陥没していたり、部屋の入口に角材が投げ込まれたままであったり、家の中で1つの部屋だけが捩れていたりする。
彼の建築にすぐに見てとれる崩壊のモチーフは、例えばSITEの仕事や、その影響を受けて「崩れた壁」を華麗な装飾的要素として取り込んだ若い世代による商業建築などとは、対極的である。それは都市の世紀末風景でもない。建築を構想すること、建設することにつきまとって離れない強迫観念として、吉田の建築を現実的に裏打ちしているものである。
彼の建築は、彼自身とスタッフにより手づくりで成り立っている部分が多い。安部公房の「砂の女」のように、手と肌に触れてくる素材のなかに、可能な建設の営みを見せている。その出来上がったものを見る限りは、吉田は建築家であることを避けてアーティストに向かうように見えるが、現場ではもっとも建築家であるべき責務に応えているのである。

■この住宅が、どれほど特異な敷地とその環境のなかに建てられたかは、サイト・プランを見ただけで十分納得できる。古い木造住宅が半ば朽ちたままにひしめいて、道の脈絡も判然としなくなった、その不整形の隙間に、鍥のように打ち込まれた、2階建てコンクリートの住宅。
鉄筋や型枠などの建設資材は、レッカーで隣家の屋根越しに搬入し、露地の隙間からコンクリートを流し込んだ。軀体施工以降の工事は、すべて吉田とスタッフによる。アドリブ的なデザインもそれで可能になった。
幅1m足らずの道路に面して入口があり、上がったところが玄関のドア。ドアの下半分はコンクリートが詰められている。
入口を入った三角形のスペースが台所・食堂・居間。それに、演劇舞台のような和室がコンクリートの壁に挿入されている。
もうひとつ奥の三角形のスペースが寝室。敷地の形から取り出した一番長い壁に沿った通路が、2つのスペースを結ぶ。四囲の家が迫っているために、開口部は最小限、ハイサイド・ライトと床に近い通風孔であり、現在のところ、生活は2階で完結させ、1階は地下室のような予備室の扱いとなっている。入口の階段の横に1階への入口があり、寝室の一番奥にも、1階へ降りるタラップがある。まだ工事も生活も、多くの部分が残されている。

1 玄関まわり。下のスケッチに見られるように、崩れた壁のイメージが反復されたドアが吊られている。右手の格子のドアを開けると階段があり、2階の玄関に達している。2階から和室の壁が突き出ている。
2 アナーキーな環境のなかで、長大なコンクリートの壁にとられた開口部は、採光と通風の効果を期待されている。
3 2階。和室。照明器具と化した紙の天井は、吉田がよく用いる手法である。

1

1 Entrance. The door having a number of repeated images of a wall half destroyed is suspended, as you can see it in the sketch on the left, if you open the lattice door on your right side, you can see a flight of stairs which reaches a vestibule on the second floor. A wall of the Japanese-style room is projecting from the second floor.
2 The opening on the huge concrete wall is expected the effects of lighting and ventilation, in this anarchic circumstances.
3 The Japanese style room on the second floor. The ceiling of paper utilized also for lightings is often fo[illegible]d in the technique of construction designed by Mr. Yoshida.

2

3

4

■The ferroconcrete designed by Mr. Yasuo Yoshida is a vision of sand...flowing sand and eternally breaking sand.
He is an architect who actually uses sand letting the customers' shoes turn white. When he was involved with the stage art for theatrical performances, he used a method to continuously drop sand from the ceiling like a waterfall. The goal of his architecture is to express the breaking, rusting and cracking of concrete, iron, paper and cloth. The room to room combinations shows unstable appearances; one part of an adjoining room is abruptly depressed, or rectangular lumber is scattered at the entrance of a room, or one room of the whole house is distorted.
This collapsing motif easily observable in his architecture, is in contrast with commercial architecture by the younger generation who introduce "collapsing walls" into their ornamental expression. Mr. Yoshida's motifs are not the spectacles of the town nearing the end of the century. They back Mr. Yoshida's architecture in a realistic sense as an obsession that sticks to shaping and building structures.
A great portion of his architecture is actually made by him and his staff manually. As in the novel by Mr. Kobo Abe "A Woman of Sand", his architecture displays the working of architecture in raw materials which can be felt by hand and skin. From the appearance of his works, it seems that Mr. Yoshida is trying to avoid being an architect to be an artist instead. But he still manages the responsibilities imposed on a true architect at the site.

■It is clear with a glance at the site plan that this house was built on a peculiar site and its peculiar environment. Old, partially decayed wooden houses jostle each other, and roads are disconnected and disordered. A concrete two-story house was dropped into a formless gap. Construction materials — such as iron reinforcing bars and molding frames — were delivered by a wrecker over the roof of the adjoining house and concrete was pushed through a gap.
All the work after the dismantlement were done by Mr. Yoshida and his staff, which made it possible for them to apply ad-lib designs.
There is an entrance facing the road of less than one meter in width. When the hanging door in the half-broken iron-barred gate is opened, there is a stairway leading directly to the second floor with the front door at the top of the stairway. The lower half of the front door is concrete. In a triangular space at the entrance, there are a kitchen, dining room and living room. Additionally, a Japanese style room is inserted like a theorètical stage into the concrete wall.
Another triangular space in the back is a bedroom. The two spaces are connected by a passage lying along the longest edge of the site. Because the houses is closed in on all four sides, openings are minimized, with only high sidelights, and ventilation holes near the floor. At present, all living is done on the second floor, and the first floor is used as a reserve room like a cellar. There is an entrance leading to the first floor next to the stairway at the entrance, and there is also a movable ladder going down to the first floor at the far back of the bedroom. Many parts of the construction are yet incomplete.

4 無秩序に密集した家々のなかに建つ。建物の形が、敷地の形状をそのまま表している。写真右下の部分が入口。
5 6 7 幅1m足らずに路地に面した玄関まわり。鉄の格子とコンクリートのドアが、1本のH型鋼から吊られている。崩壊のイメージは、この2つのドアや郵便受けのサインに反復されている。階段をあがったところにある玄関ドアも、ガラスと異形鉄筋とコンクリートで、同様のイメージを詩的に、荒々しく繰り返している。

4 The form of the residence which stands surrounded by lots of houses tightly crowding in disorder represents itself the shape of the site. The entrance is shown at the lower part on the right in the picture.
5 6 7 The foyer area facing the alley is less than one meter in width. An iron grid door and another concrete door are suspended from an H-beam. The image of disruption is repeated in these two doors and in the signature on the mailbox. The entrance door at the top of the stairs also repeats a similar image roughly and poetically in its glass, deformed bars and concrete.

5

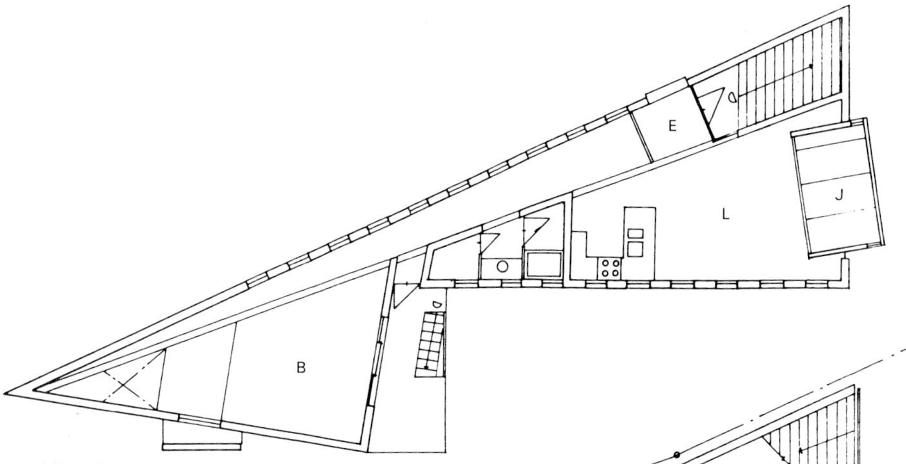

Second floor plan

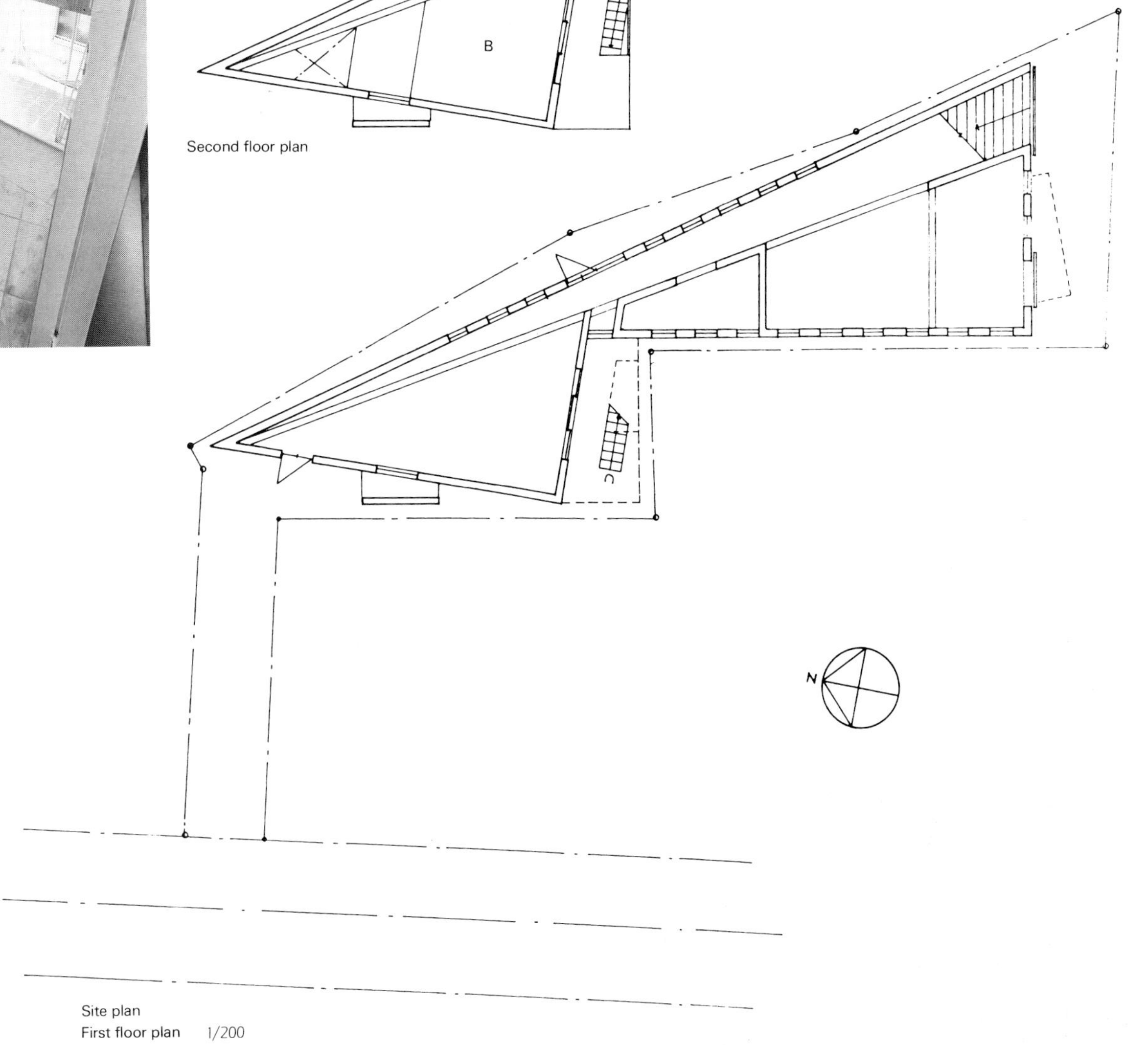

Site plan
First floor plan 1/200

6

7

8

8 2階。玄関を入ったところが、居間と台所。コンクリートの長い壁に沿った廊下の奥に寝室がある。
9 10 1階の一番奥に、2階の寝室に上がる鉄の梯子がある。この部分だけが吹き抜けている。1階は全て予備室として残されたままになっている。
11 玄関わきのコンクリートの吊り戸を開けて内部に入ったところは、まるで地下室のような印象。黙示録的な演劇の舞台とも見える。

8 The second floor. The sitting room and the kitchen follow the vestibule. You will be shown into the bedroom after passing through the corridor along the long walls.
9 10 You can find a flight of stairs of iron to the bedroom upstairs at the end of the ground floor. Only this part is void. They leave preliminary all the rooms on the first floor.
11 The inner part of vestibule will give you a strange impression as if you entered a sub-ground, when you open the suspended door. It looks like an apocalypticscene of stage.

9

10

11

吉田邸 [Yoshida House]

■大阪のベッドタウン、といっても建売住宅が軒を接して並ぶ、下町的な環境の一画に建つ3階のコンクリート住宅。
四隅に、玄関、階段室、便所、収納などのコアを立ち上げ、そこに解放的な床をかけている。1階に、台所、食堂、居間、2階に主寝室、3階に子供室が想定されているが、まだ未確定の部分が多く残されている。というのは、着工以来ほぼ4年目を迎えるこの建物は、吉田とそのスタッフによる自力建設が、まだ続いているからだ。
吉田がこれまで手がけた建築の多くも手づくり的だった。しかし、軀体工事は工務店が行い、手摺やドア、床や壁仕上げの一部、家具、照明器具などの、限られた部分において、吉田とスタッフの手が加えられていたわけだが、この住宅においては、基礎工事から最終仕上げまでの全工程を、自分たちの手で体験するという試みが貫かれている。
同時に、ここでは施工上の合理性が、過激なまでに目論まれている。1、2階までのコンクリートの仮枠はガラス(!)を用い、使用後はそのまま開口部に再利用されている。鉄骨の工事足場は手摺として残される。
造形上の遊びを一切排除したという結果が、奇想ともいえる形態を現しつつある。自力建設という、狂気のような意志の力が、一見明快な構成のコンクリート建築の随所に噴出している。

■A three-floor concrete dwelling house in one of the bedroom communities of Osaka, where pre-built houses stand side by side creating a rather "Shitamachi" atmosphere. The foyer, staircase, toilet, and closet are raised up from the four corners, and an open floor is placed above them. Kitchen, dining room, and living room on the first floor, master bedroom on the second floor, and children's room on the third floor — this arrangement is still in the imaginary stage, much has yet to be decided. It has been almost four years since construction began, and construction by Mr. Yoshida and his staff is still to be continued.
Many of the buildings Mr. Yoshida has built so far were largely handmade. Usually, the basic body work is done by a building contractor, then limited portions such as handrails, doors, parts of the floor, wall partitions, furniture and lighting equipment, are done by Mr. Yoshida and his staff. In this building, they tried to do all construction work from the foundation work to final finishing.
The plan is very rational. The temporary glass frames of the concrete, extending from the first floor to the second floor, remain as they are and are reused as openings after use. The steel-frame scaffolding will be used as handrails.
A complete removal of refinement in the constructive art results in a fanciful style. Their desire to build this by themselves shows everywhere in the concrete building.

123 コンクリートの軀体は打ち上がり、床スラブもかけ終えている。厨房機器も入った段階。サッシュを完全にし、玄関まわりと水まわりを整備して、電気・ガス工事ができれば、もう住むことができるだろう。

123 The concrete base structure is completed and floor slabs are boarded. Kitchen utensils and appliances are already installed. It is possible to start living in the house when it is sashed completely, the entrance area and water-supply facilities are operable and when electrical and gas work is completed.

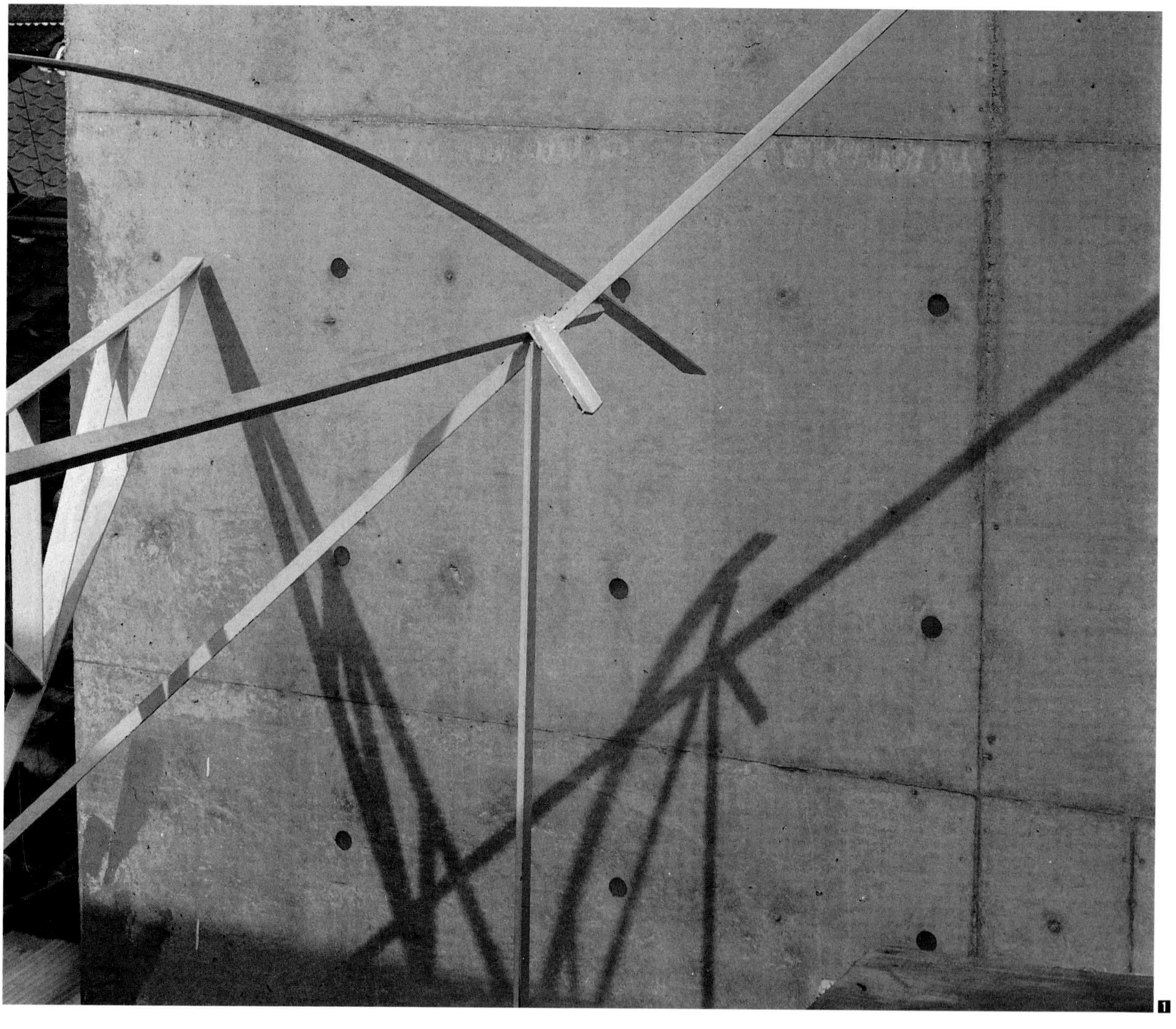
1

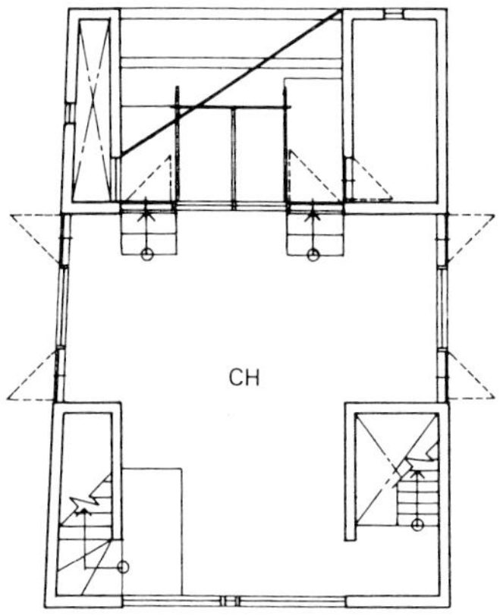

Third floor plan

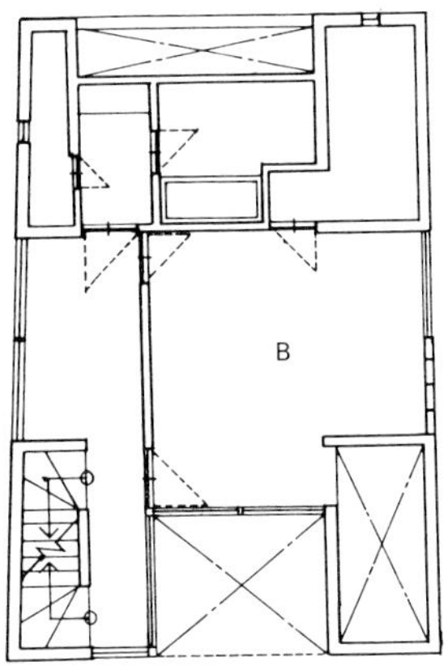

Second floor plan

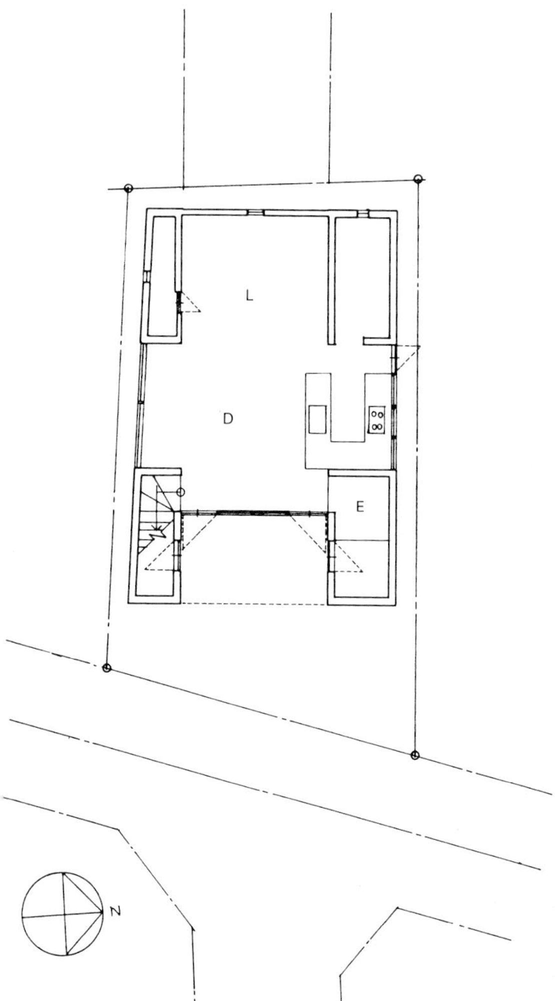

1/200
Site plan
First floor plan

2

3

4

4配置図を見ればわかるように、十字路のところで立往生してしまったような敷地に建っている。東から歩いてくると、T字型の道の正面にファサードが見えるし、写真のように西から見ると、建売住宅が両側に並ぶ袋小路の突き当たりに、足場を組んだままの巨大なコンクリートのマッスが聳えている。しかし、手づくりの柔らかく、気さくな表情が至るところに現れていて、子供たちの多いこの街区に、意外といっていいほど馴染んでいる。

4As seen from site plan drawing, the house looks stuck at a crossroads. When coming from the east side, the facade is seen in front of the T-intersection. When viewed from the west, as in the photograph, a huge concrete mass with scaffolding rises high at the end of a dead alley where pre-built houses line up on its west. All parts of the house have a hand-made softness and casual expression. It is surprisingly fitted to this district with a high population of small children.

建築概要［Building Outline］

作家紹介［Biographies］

あとがき［Afterword］

建築概要 [Building Outline]

●凡例[注：データーは各事務所の公表数値に準じた]

※ No.6 a, b and c indicate plot area, building area and total floor area, respectively.

1. 所在地
2. 施主
3. 家族構成
4. 設計者　a. 建築　b. 構造　c. 設備
5. 施工者
6. 面　積　a. 敷地面積　b. 建築面積　c. 延床面積
7. 構　造
8. 外部仕上げ
9. 内部仕上げ
10. 使用機器
11. 打放しコンクリートに関しての型枠材料、サイズ、設計、施工上の狙い。コンクリートの質処理、表面処理。他材料とのとり合せ上の狙い。ディテール上の処理についての作家コメント。

阿部 勤 [ABE Tsutomu] 16▶25

臼杵邸 [Usuki House]

1. 東京都練馬区
2. 臼杵洋三
3. 夫婦＋子供２人
4. a. 阿部勤/アルテック建築研究所　b. 森勇建築構造設計事務所　c. アルテック建築研究所
5. 東興建築
6. a. 259.956㎡　b. 82.274㎡　c. 190.218㎡
7. 地下１階・地上１階：RC造、２階：木造
8. RC、木造部分：防火サイディング
9. RC・椙ベニア目透し貼
11. 私の家の延長上にあるもので、１階をコンクリート打放しの壁構造とし、２階を木造軸組構造とし木造の屋根で被う混構造となっている。
   ●普通打放し型枠600×1,800
   ●打放しの表面の素材感を生かすため補修は最小限に押え、撥水剤を塗布。
   ●塗壁・木等で仕上がったアルコーブや個室とのコントラストを狙う。

綿貫飯田邸 [Watanuki-Iida House]

1. 東京都杉並区
2. 綿貫建太郎・飯田隆
3. 両親＋夫婦＋子供２人
4. a. 阿部勤アルテック建築研究所　b. 森勇建築構造設計事務所　c. アルテック建築研究所
5. ミトモ建設株式会社
6. a. 161.5㎡　b. 64.5㎡　c. 129㎡
7. 地下１・地上１・２階 壁：RC壁構造、小屋・屋根裏：鉄骨＋木造
8. RC、サイディング
9. RC、ペンキ仕上げ
11. １階に両親の住いと薬局、２階にコンクリート打放しの壁の空間、それを被う屋根という型で構成。
   ①普通打放し型枠
   ②撥水剤塗布
   ③白く塗装した木部との組合せ
   ④居室部は内部打放し仕上げとし外に断熱材の上サイティンク仕上げ、打放しコンクリート面保護のための庇、水切り。

安藤忠雄 [ANDO Tadao] 26▶47

細工谷の家 [Town House in Saikudani]

1. 大阪市天王寺区
2. 野口幸一郎
3. 夫婦＋子供２人＋老夫婦
4. a. 安藤忠雄建築研究所　b. ㈱アスコラル構造研究所
5. 建築：㈱久栄工務店、設備：関西工業、電気：宮本電設、家具：守谷建具
6. a. 68.45㎡　b. 40.00㎡　c. 106.26㎡
7. 鉄筋コンクリート壁式構造
8. 外壁：ベニヤ型枠RC、防水剤塗布　屋根：コンクリートスラブの上、シート防水砂利敷 床：玄晶石貼角300、厚10
9. 天井：ベニヤ型枠RC　内壁：ベニヤ型枠RC　床：玄晶石貼角300厚10、一部　木軸下地ブナフローリング乱尺貼厚15
10. 電気：電気幹線設備、電灯差込設備、照明器具設備、電話、インターホン設備　衛生：給排水設備、給湯設備　衛生器具設備　空調：換気設備、冷暖房設備

孫邸 [Son House]

1. 大阪市天王寺区
2. 孫世光
3. 夫婦＋子供３人
4. a. 安藤忠雄建築研究所　b. ㈱アスコラル構造研究所
5. 建築：㈱久栄工務店、設備：城戸工業、電気：宮本電設、家具：守谷建具
6. a. 103.3㎡　b. 85.15㎡　c. 206.51㎡
7. 鉄筋コンクリート壁式構造
8. 外壁：ベニヤ型枠RC、防水剤塗布　屋根：コンクリートスラブの上シート防水　床：玄晶石貼300×600厚20
9. 天井：ベニヤ型枠RC　内壁：ベニヤ型枠RC　床：木軸下地ブナフローリング乱尺貼厚15
10. 電気：電気幹線設備、電灯差込設備、照明器具設備、電話、インターホン設備　衛生：給排水設備、給湯設備、衛生器具設備　空調：換気設備、冷暖房設備

畑邸[Hata House]

1. 兵庫県西宮市
2. 畑眞八郎
3. 夫婦＋子供１人
4. a. 安藤忠雄建築研究所　b. ㈱アスコラル構造研究所
5. 建築：大工建設株式会社　設備：㈱山口商会　電気：神建産業㈱　家具：マツヤ家具
6. a. 441.49㎡　b. 118.73㎡　c. 207.20㎡
7. 鉄筋コンクリート壁式構造
8. 外壁：ベニヤ型枠RC　屋根：コンクリートスラブの上アスファルト防水、砂利敷
9. 天井：ベニヤ型枠RC　内壁：ベニヤ型枠RC　床：木軸下地ブナフローリング乱尺貼厚15　一部　玄昌石貼300×600厚20、木軸下地ココヤシカーペット貼厚9.木軸下地タタミ敷
10. 電気：電気幹線設備、電灯差込設備、照明器具設備、電話、インターホン設備、テレビ共聴設備　衛生：給排水設備、給湯設備、プロパンガス設備、衛生器具設備　空調：換気設備、冷暖房設備
11. 無機的なコンクリートの壁と、そこに作用する光は、私の建築に対するイメージの、原点である。規則正しく取り付けられた、コンクリート仮枠パネルとセパレーターの痕跡。等質に仕上げられた表面と鋭いエッジ。しかし、コルビュジエ後期の作品に見られるような、素材を強調した、なまなましい彫塑的空間は好みとしない。むしろ、力強さを背後に秘めた、ひそやかな、無機的な素材として、コンクリートを語らせたい。そして、それがつくり出す、この冷ややかで静かな空間に、光を引き込むことによって、素材の意志を超えた、柔らかさと透明感のある空間へと解き放ちたい。

東 孝光[AZUMA Takamitsu] 48▶59

登美ヶ丘の家[Tomi House]

1. 奈良市西登美ヶ丘
2. 玉垣嘉久
3. 夫婦＋子供１人
4. a. 東孝光＋東環境・建築研究所　b. 構造計画研究所　c. 東孝光＋東環境・建築研究所
5. ㈱秦工務店
6. a. 440.0㎡　b. 153.7㎡　c. 160.6㎡
7. RC造平屋
8. RC
9. RC
10. 暖・冷房：ダイキン空冷ヒートポンプ式エアコン　衛生機器：TOTO

遊ハウス[YU House]

1. 東京都新宿区
2. 吉川 透
3. 夫婦
4. a. 東孝光＋東環境・建築研究所　b. 構造計画研究所　c. 東孝光＋東環境・建築研究所
5. ㈱篠崎工務店
6. a. 71.2㎡　b. 42.6㎡　c. 101.38㎡
7. RC造　地下１階地上２階
8. RC
9. RC
10. 暖・冷房：三菱クリーンヒーターエアコン　衛生機器：TOTO
11. 石灰岩を加工してつくったセメントに、砂利と砂を混ぜて水で練り、型枠に流し込んで固めるコンクリートを、私は文字どおり地球の大地からの賜り物としてこよなく愛している。その力学的な長所や自由に形をつくれる可塑性とともに、人間の手のわざの痕跡や、四季の変化や光と影の移ろいを細やかに映し出してくれる。コンクリート打放し。それは私たちの祖先が永らく木や竹や紙や土という自然から引き出した素材の美しさにつながるものであり、日本人の資質にも合った素材として、有力な建築空間の表現の可能性をもつと私は考えている。

原 広司[HARA Hiroshi] 60▶77

夢舞台[The Stage of Dreams]

1. 静岡県伊東市
2. 中塚大輔
3. 夫婦＋子供３人
4. a. 原広司＋アトリエ・ファイ建築研究所　b. 佐野建築構造事務所
5. 三海建設
6. a. 455.15㎡　b. 90.99㎡　c. 145.80㎡
7. RC一部木造
8. 屋根：カラー鉄板瓦棒葺き　外壁：石綿スレートVP、ステンレス板貼、RC化粧型枠
9. 床：ジュータン、ラワンベニヤoscl　壁：桜ベニヤ目透し貼VP天井：プラッターボード下地、寒冷紗貼VP

虔十公園林フォリストハウス[Kenju Park Foreset House]

1. 宮城県加美郡
2. 小野裕子
3. 夫婦＋若夫婦＋子供３人
4. a. 原広司＋アトリエ・ファイ建築研究所　b. 佐野建築構造事務所
5. 木戸建設
6. a. 5,870㎡　b. 254.89㎡　c. 355.61㎡
7. 鉄筋コンクリートおよび木造
8. 屋根：アスファルト防水玉砂利押え　カラー鉄板平葺き、一部瓦棒葺き　外壁：磁器質系45角タイル、RC化粧型枠、フレキシブル板VP
9. 床：ラワンベニヤoscl、テラゾータイル、ブナフローリングブロック　壁：プラスターボード下地、ビニールクロス貼、桜合板目透し貼VP　天井：プラスターボード下地、ビニールクロス貼
10. システムキッチン、空冷式ヒートポンプエアコン、灯油FF暖房器、ガスFF暖房器
11. さまざまな材質を寄せ集めるとなると、コンクリートの打放し面は貴重なマチエールとなる。特に、モノクロームな感じの室内であると、金属板や塗装あるいはクロス貼りの均質の面から、石貼りの模様のある面にいたるグラデーションのなかで、打放しの面は中間に位置するので、必ずしも全面的に活用しなくとも意味をもってくる。〈夢舞台〉や〈フォリストハウス〉での打放し面は、そのようなマチエールを混ぜる表現のなかのひとつの要素である。この活用法は、室内に限らず建物の外装に対しても応用できる。一般に、マチエールの寄せ集めの一要素とするときには、打放し面特有の構築性は消える。ただ、柱や梁の打放し面では力学的な意味での構築性は残るから、単なるマチエールの集合をこえたブリコラージュの面白さがだせるように思える。

長谷川逸子[HASEGAWA Itsuko] 78▶87

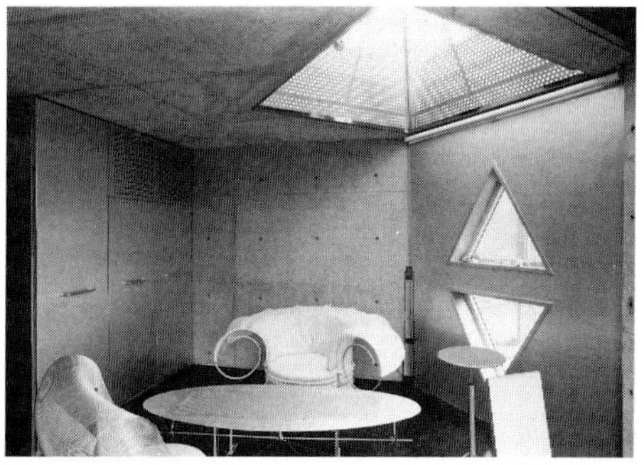

NCハウス[NC House]

1. 東京都中野区
2. 個人
3. 共同住宅
4. a. 長谷川逸子建築計画工房　b. 服部重信　c. 長谷川逸子建築計画工房
5. 大成建設東京支店
6. a. 269.11㎡　b. 160.24㎡　c. 430.20㎡(1F141.13㎡、2F141.13㎡、3F139.30㎡、RF8.64㎡)
7. 鉄筋コンクリート造
8. RCビニタイル塗布(一部メタラックC)、アルミパンチングメタル張
9. 床：コルクタイル貼　天井：RC壁：RCおよび桐ベニヤVP、一部アルポリック張
10. 電気：低圧受電　衛生：受水槽加圧ポンプ方式、ガス焚瞬間湯沸器　空調：暖房ガス焚ファンコイル方式　冷房セパレート型ヒートポンプ方式
11. 内と外の境界をつくる材料が何であるかは、その建築のあり方を大きく左右する。境界は硬く粗くそして厚みのある壁、つまりコンクリートのようなものより、外気と融合するほど薄く軽くこまやかさを備えた膜のようなものでありたい。そのほうが私達の風土や神経に合うと私は思っている。このNCハウスは設計条件でやむなくコンクリート造となったがそのコンクリートにもう一重アルミパンチングメタルという膜を張りめぐらしてある。

早川邦彦[HAYAKAWA Kunihiko] 88▶101

STRATUM

1. 東京都中野区
4. a. 早川邦彦建築研究室　b. 百田構造設計　c. 早川邦彦建築研究室
5. 大栄工務店(彫刻：山本衛士)
6. a. 65.72㎡　b. 45.99㎡　c. 131.44㎡
7. 鉄筋コンクリート壁式構造
8. RCアクアシール塗布、コンクリートの上アクリルリシン吹付け／VP／ボンフロン吹付け、アルミニウム菱葺き
9. 床：パンチカーペット　内壁：PB厚12寒冷紗AEP、RC、コンクリートの上アクリルリシン吹付け　天井：PB厚9寒冷紗AEP
10. キッチンセット(サンウェーヴ)、ミニキッチン(アトラス)、ユニットバス(TOTO)、エアコン(ダイキン)、照明(ヤマギワ)
11. "ストラタム"は、1階を店舗、2階をオフィス・スペース、3・4階はメゾネット・アパートという、異なる機能が相互に積み重ねられた構成をとっている。周辺の住居のスケールである2層分の高さをもつコンクリート打放しの壁は、平行四辺形の敷地境界線をほぼ忠実にトレースして、過密な外側の世界から、内側の世界をかこいとる機能をもつと同時に、異なる色彩と形態が積層する全体の基部ともなっている。

STEPS

1. 東京都国立市
4. 早川邦彦建築研究室
5. 大栄工務店
6. a. 224.80㎡　b. 173.04㎡　c. 449.59㎡
7. 鉄筋コンクリート壁式構造
8. RCアクアシール塗布、一部コンクリートアクリルリシン吹付、VP
9. 床：パンチカーペット、壁：PB厚12寒冷紗AEP、天井：PB厚9寒冷紗AEP
10. システムキッチン(サンウェーヴ)、水洗器具(TOTO)、エアコン(ダイキン)、浴室器具(TOTO)、ユニットバス(TOTO)
11. "ステップス"は、8戸の賃貸住戸を中心とし、1階の店舗と屋上を利用したスカイ・ミュージアムにより構成される。階段群、屋上のスカイ・ミュージアムのゲート、そして空中を走るタイコ橋などを垂直方向に積み上げ、共有領域に演劇的な空間を導入するよう意図した。
また、階段群を挟む2枚のコンクリート打放し壁の内側の彩色された面には、それぞれの住戸の異なった形状の開口が取り付けられ、各住居の内側で展開される生活の気配が適度に共有領域に発信される。

伊東豊雄[ITO Toyo] 102▶109

馬込沢の家[House in Magomezawa]

1. 千葉県船橋市
4. 伊東豊雄建築設計事務所
5. バウ建設㈱：建石実
6. a. 100.00㎡　b. 49.70㎡　c. 81.20㎡
7. RC造＋鉄骨造
8. 屋根：カラー鉄板厚0.4平葺、外壁：コンクリート打放し、石綿パーライトボード厚6ケミストップ塗布
9. 床：モルタル金ゴテ押え　ラミンフローリング
10. クリーンヒーターエアコン(灯油)
11. 1階と2階に1部屋ずつの、小さな家である。2階がオープンなスペース、1階がクローズされたスペースで、我々が考えてきた、2つの対照的なスペースを使い分けながら都市に住むというテーマを、最も単純に表現している。1階は、コンクリートの壁に囲まれ、やや掘り込まれた床をもつ箱状のスペースで、このコンクリートボックスは地面の延長であるとも、また2階の鉄骨を支える基壇であるともいえよう。外部だけでなく内部にも打放しコンクリートが用いられているが、これは打放しコンクリートが特別な意味や思い入れを発しない、中性的な存在として据えられているからである。コンクリートの6m×6mという平面は、壁式構造では最大に近いスパンであるが、スラブ厚を200mmのダブル配筋とすることで、壁厚は150mmのまま、中央部に梁を設けることなく成立している。

石井和紘[ISHII Kazuhiro] 110▶117

反ブルネレスキ・ハウス[Anti-Brunelleschi House]

1. 神奈川県鎌倉市
2. 細野孝興
3. 2世帯(老夫婦+夫婦+子供2人)
4. 石井和紘建築研究所
5. 第一建設工業
6. a. 298.90㎡ b. 103.68㎡ c. 214.97㎡
7. 鉄筋コンクリート造
8. 屋根:シート防水の上客土敷込み芝張り 外壁:鉄筋コンクリートおよびメースの上シーラー処理AEP拭き取りクリアー吹付け 建具:アルミサッシュアルマイト仕上げ
9. 床:ラミンフローリング厚12CL 壁:シーラ処理AEP拭取り 天井:シーラ処理AEP拭取り
10. 暖・冷房:ガスFFクリーンヒーター
11. 反ブルネレスキ・ハウスは、パラッツォ・ドゥカーレのコーナーを、建築家がそう考えたように、本当にスリットにしようとしたものである。そしてパラッツォ・ドゥカーレでピンピンで接触したふたつの柱に、面をとって幅5cmほどのガラスを嵌めた。このことがコーナーの暗さをずいぶんと救った。さらにこのことは凹のコーナーだけでなく凸のコーナーでも同じこととなる。そして、さらに4つの面が独立していると同時に、円環的な働きももっている。つまり、反ブルネレスキの方法のなかでブルネレスキをふたたび取り込もうとした。それを可能にすることは、反ブルネレスキもブルネレスキと同じに、一般柱が1本柱になっているところにもう一度スリットを入れることであった。これによって各柱とコーナーの柱は、このスリットの扱いにおいて全く同じになったのである。
    自分としてはこれで反ブルネレスキ側のブルネレスキができたと思ったのが、それを表現するためにもうひとつ処理した。パラッツォ・ドゥカーレでコーナーにおいて横の連続梁(文字入り)が角柱で切られているのを止めたのである。つまり、横梁はぐるりと回ることで、必ずしもこれが正方形の中庭でなくてもよい、つまり円形でも、角形でもよい。しかも、ブルネレスキを取り込んだ反ブルネレスキに到達したように思えたのである。それがこの30cm角柱間の30cmスリットの説明である。さらに耐力壁なしの構造であるので、外周(非中庭面)は横力補強をするために中をつないでH型にして2本の柱が一体に効くようにした。

石井 修[ISHII Osamu] 118▶129

目神山の家8[House in Mekamiyama 8]

1. 兵庫県西宮市
2. 鈴木松吉
3. 夫婦+子供1人
4. a. 石井修 b. ㈱キンキ構造設計 c. 美建・設計事務所
5. 建築:モリタ建設㈱ 設備:㈱竹中工業所 電気:河村電気㈱
6. a. 341.38㎡ b. 83.89㎡ c. 202.81㎡
7. 鉄筋コンクリート造
8. 屋根:アスファルト防水、砂利敷きの上肥土 外壁:コンクリート打放しの上、ケミストップ
9. 床:ベラミカセラフロア—200角貼、ヒューガカーペト敷、コットンカーペット敷 壁:RCの上APクリア、桜ベニヤ張りの上、クリアラッカー、白レンガ半枚積 天井:RCの上APクリア、ラワンベニア下地クロス貼
11. ①樹脂合板900×1,800横使いでパネル割りおよび木コンの位置図を作成した。
    ②ギャラリーは間口が狭く、床はタイル貼で壁、天井はコンクリート打放しなので、天井面にフィルムケース(約2,000個)をT型に組み合せて打ち込んだ。
    ③かた練りのコンクリートを使い、コンクリート打ちは、所員数名も参加して十分コンクリートつつきを行った。
    ④膨張性セメント混和剤、デンカCSAを用いた。
    ⑤ホールの階段横に設けられた飾台をかねた手すりは、コンクリート打放しで3mのキャンティレバーになっているので、PC工法を用いている。
    ⑥室内はすべてをコンクリート打放しとせず貼と使いわけた。

光明台の家[House in Komyodai]

1. 大阪府和泉市
2. 新田剛
3. 夫婦
4. a. 石井修 b. ㈱キンキ構造設計 c. 美建・設計事務所
5. 建築:㈾中野工務店 電気:河村電気㈱ 設備:㈱竹中工業所
6. a. 271.44㎡ b. 105.99㎡ c. 154.22㎡
7. 1階:鉄筋コンクリート壁式構造 2階:木造
8. 外壁:RC、一部杉板横貼XD塗 屋根:波型スレート葺
9. 床:杉板貼XD塗 壁:RC打放し 天井:カラ松板横貼XD塗
11. ①樹脂塗型枠900×1,800横使いとし、ファサード面の幾何学的な形態に呼応して現れるシンメトリーなコンパネ割が建物全体にも及んでいる。
    ②膨脹性セメント混和材を生コンに混入し、RC打放し面の緻密さをだした。
    ③杉板、カラ松板、スレート、和紙、土壁等の仕上材料の素材そのものを生かす意味でRC打放しも同等に考え、これらの対比や融合を試みた。
    ④RC打放しを強調するため、出隅の面取りは設けず、窓枠廻り等にも他の部材ができるだけ現れないディテールを採用した。また、玄関アプローチのRC出隅には意図的に化粧面を設け、他とは異なる表情を与えている。

槇 文彦[MAKI Fumihiko] 130▶137

自邸[Maki's House]

1. 東京都品川区
2. 槇文彦
3. 5人
4. a. 槇文彦
5. 東興建設
6. a. 318.77㎡ b. 158.38㎡ c. 254.30㎡
7. RC造
8. 外壁：RC塗装
9. 床：じゅうたん 壁：石こうボード、モルタル下地EP
天井：石こうボードEP
11. 設計上の狙い
住宅の外壁として、板の木目がみえるようなコンクリート打放し面をつくりたいが、打放しはとくに都市部ではよごれが著しい。塗装によって、ある程度の耐候性をもたせたいがその場合、型枠の木目やコンクリートの素材感を損なわないような塗装の手法を考えた。
型枠：材料サイズ等
杉板、厚さ15mm、3種類の板幅(150、180、210mm)のものを乱に割りつけた。また、板と板との継ぎ目は突きつけにして、板の反り等による自然な目ちがいがでやすいようにした。
コンクリート表面処理
防水シーラーで下地処理を行い、稀めのアクリルエナメルを刷毛塗りしながら、杉板の木目が浮き出るように、1枚1枚拭きとりを行う。コンクリートの地肌の色ちがいも塗装で調整しながら全体のトーンを合せていく。最後に保護層として、アクリルクリヤーを2回塗布。
施工の留意点
型枠精度は極力上げるが、脱型後、目ちがいやバリ等かあっても補修しないで、できるだけそのままの状態を保つように心がけた。

村上 徹[MURAKAMI Toru] 138▶143

田方の家[House in Takata]

1. 広島県広島市
2. 森木健二
3. 夫婦＋子供2人
4. 村上徹建築設計事務所(村上徹・西宮善幸)
5. 野村建設(吉村正和)
6. a. 101.11㎡ b. 34.02㎡ c. 102.06㎡
7. 鉄筋コンクリート壁式構造＋鉄骨造(屋根)
8. 外壁：合板型枠RC防水撥水剤塗 屋根：ステンレス厚0.3ー文字葺
9. 床：コンクリート直押上の上コルクタイル貼 壁・天井：合板型枠RC、プラスターボード寒冷紗VP
10. 洗面：TOTO L-306 冷暖房：ダイキン 便器：TOTO C-710 照明器具：ヤマギワ 浴槽：TOTO FBS-1104
11. 型枠材料：塗装合板型枠(900×1,800)厚12、丸柱・曲面壁は同等塗装のペンド型枠
コンクリート：普通コンクリート、4週圧縮強度180kg/㎠、スランプ18
表面処理：(外壁)木コン部にチオコールコーキングづめ防水撥水剤3回塗り (内壁)型枠脱却のまま未処理
全開口をもつ2階分のコンクリートボックスに鉄骨屋根を架け、アルミサッシュ・設備機器等を部品的に取り付けるとこの住宅の全体は完成する。
構造軀体費を最小限度に抑えた3階建のローコスト住宅。床スラブ・立上がり壁上端等はコンクリートを直押えし、コンクリートボックスとしての意味を強調している。
建具・家具・木製階段等の取り付け部もコンクリートに細かな細工はなく、双方の部品としての精度を重視し、直接ぶつけ合うディテールを採用し、構成を単純化している。

室伏次郎[MUROFUSHI Jiro] 144▶157

西原の家[House in Nishihara]

1. 東京都渋谷区
2. 金子義明
3. 夫婦＋子供1人
4. a. 室伏次郎／アルテック建築研究所 b. 森勇建築構造設計事務所 c. 友進
5. 建築・設備：福井工務店 電気・三栄電設
6. a. 99.18㎡ b. 59.17㎡ c. 175.81㎡
7. 鉄筋コンクリート壁式構造
8. RC
9. RC

土浦の家[House in Tsuchiura]

1. 茨城県土浦市
2. 寺田肇
3. 夫婦＋子供2人＋母親2人
4. a. 室伏次郎／アルテック建築研究所 b. 森勇建築構造設計事務所 c. アルテック建築研究所
5. 町屋工務店
6. a. 320.75㎡ b. 129.45㎡ c. 237.75㎡
7. 鉄筋コンクリート造＋外壁木構造
8. ガルマニウム鋼板貼、一部RC
9. 木造軸組のまま、一部RC
11. 2つの計画ともに打放しコンクリートは、単純な素材のみで構成された軀体であることを設計上の狙いとしたためである。コンクリートの色合い、表面の質感自体を空間の構成要素としようとするものではない。それがただ、素材のみで、基本的シェルターとしての性能をもち得るということで、コンクリートは極めて経済的な材料である。また私の場合、単一素材で構成できるということおよびその性能のみが重要であり、表面性について、デザインの操作の対象とはしていない。コンクリートの壁面として、デザインの対象とするのは、面の大きさ、空間のボリューム、開口のあり方等の相互の関係性のみを対象としている。したがって、型枠、割付、表面処理等すべて、極めて一般的な方法によっており、製作に特別な時間、労力、費用を掛けるということはない。コンクリートの壁面はすべてがその型枠の内容に支配されているが、でき上がったものは、ごく普通の表面と性能をもっていればそれでよく、特別に磨き上げたような表面をもちたいとは考えない。いささか偶然性に支配されざるを得ない材料であって、多少の荒さもまたやむを得ないものであろう。かといってその荒さを期待したという意味でその表面性にコンクリートらしさを求めるべきものとも考えない。

内藤 廣[NAITO Hiroshi] 158▶165

共生住居[The Commensally House]
1. 神奈川県鎌倉市
2. 自邸
4. a. 内藤廣建築設計事務所 b. SDG構造設計集団 c. 内藤廣建築設計事務所
5. ㈱伝建
6. a. 462㎡ b. 140㎡ c. 237㎡
7. 鉄筋コンクリート造(薄肉ラーメン構造)
8. RC
9. RC、ラワン板壁O.S
10. キッチン：キッチンハウス、湯沸器：ノーリツ、暖・冷房：ノーリツ温水暖房
11. この建物は、RCの軀体・建具・仮設的間仕切で成り立っている。そのなかで、RCをローコスト化のひとつの手段として考えた。軀体費は、通常総コストの25〜30%を占めるが、まずこれを20%近くにもっていくこと、そして、軀体打ち上がり時に天井、枠などが終っているようにディテールを構造の形式と平行して考えた。結果、天井は吸音、断熱をかねてドリゾール板打ち込み、建具枠はこれとゾロにスチール枠を打ち込んだ。また、横枠は軀体に溝を切っただけのものとした。こうすることで、天井仕上、枠工事分のコストを軀体にもたせることができた。建具と軀体との納まりを単純にするため、構造事務所と相談して、架構を壁柱、床梁の組み合せによる変形の薄肉ラーメン構造とした。RC打放し面は、型枠を地組みし、大型型枠を作って立ち上げた。試みに、剥離材にオイルを使った。このため打放し面は、トラバーチンを磨きあげたような風合いが出た。
打放しの納まりは、精度のある建具等とのとりあいが問題だが、この建物では、可能な限りいろいろな性能を打放しにもたせたため、RC工事にはきわめて高い?精度が要求された。その分、軀体と建具、間仕切の存在がはっきりときわ立ち、明解な建物になった。

六角鬼丈[ROKKAKU Kijo] 166▶177

塚田邸パートII[Tsukada House PartII]
1. 群馬県沼田市
2. 塚田恒徳
3. 夫婦(主家部分若夫婦＋子供3人)
4. a. 六角鬼丈計画工房 b. 花輪建築構造設計事務所 c. 六角鬼丈計画工房
5. ㈱大淵建設
6. a. 291.98㎡ b. 全体135.57㎡(パートII 34.57㎡) c. 全体178.89㎡(パートII 41.19㎡)
7. 鉄筋コンクリート造平屋、屋上温室鉄骨造
8. 外壁：RC 屋上：スタイロ断熱上アスファルト防水シンダーコンクリート70㎜
9. 天井：RC 床：スタイロコンパネとタモ合板貼ウレタン塗 壁：スタイロコンパネ37㎜打込みと寒冷紗貼パテVP仕上げ
10. FF暖房器
11. ①型枠 ラワン・ベニア 3'×6' パネコート処理(剥離剤)経済性
②スランプ18表面処理なし
③木建とアクリルのフィックス窓
はき出しの木建は足元の防腐に対しては塗膜防水をたちあげ、さらに接点にコーキングという二重防御をしている。アクリル部は曲線を描くため、サッシ(高価)をなしとし、コンクリート目地にアクリル板をフィックスコーキング止めとし、継ぎ部のみにSUSのフラットバーを使用した。

青葉台の家[House in Aobadai]
1. 神奈川県横浜市
2. 花輪紀昭
3. 夫婦＋子供2人
4. a. 六角鬼丈計画工房 b. 花輪建築構造設計事務所 c. 機械設備：フジゴウ技術士事務所 電気設備：松本電気設計事務所
5. 門脇建設㈱
6. a. 181.80㎡ b. 88.45㎡ c. 144.41㎡
7. 鉄筋コンクリート造2階建
8. 外壁：RC上ウエテキシ塗 サッシ スチールビニトップ塗 屋根：コンクリート上断熱クロロプレンゴム系塗膜防水
9. 天井：RC、床：木組パーライト充塡コンパネ上コルタイル貼 壁：スタイロコンパネ打込み、寒冷紗貼パテVP、一部RC
10. FF暖房＋冷暖房器(電気)
11. ①型枠 ラワン3'×6'、パネコート処理(剥離剤)、コンプラスト＋ウエテキシ塗布(クラック防止、コーティングのため)
②スランプ15
③外部は屋根を外断熱として、インテリアの天井を打放しとした。屋根が曲面のため軟質のウレタンフォームを断熱材として利用。
サッシはすべてスチールサッシ、面おさめコーキング処理のためサッシの塗装を強化した。
コンクリート打放しのコーナーはすべてピン角。

鈴木 恂[SUZUKI Makoto] 178▶183

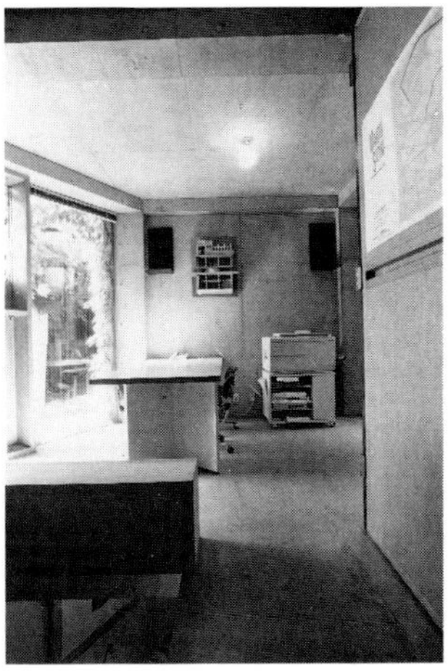

ATL 8110

1. 東京都渋谷区
2. 鈴木恂建築研究所
3. 所長＋所員４人
4. a. 鈴木恂建築研究所(担当：鈴木恂・五味睦雄) b. 佐々木睦朗構造計画研究所
5. 大栄総業
6. a. 35㎡ b. 20㎡ c. 59㎡
7. RC造地上２階 地下１階建
8. RCの上、ウォーターシール３回塗り
9. RC
10. 個別ヒートポンプ式冷暖房(ダイキン)
11. アクリル塗装型枠3×6サイズ 厚12mm ヨコ使い RC仕上げ 浸透性コンクリート防水材ウォーターシール3回塗り。住居部分の分裂とでもいうべきか。ここで提案したいことのひとつは、住宅の解体についての実験的な方法であった。生活にそれらを繋いでいくポテンシャルを保てるならば、住宅の部分が分離分散して、街のあちこちに配されることも可能なのであろうという予測。したがって、この小建築には、住居の一構成要素としての機能的なまとまりの表情と、住居の断片としての非整合な未完結な姿とが重なって表れている。住居の中でも創造的場の部分ということで、それをアトリエと呼んではいるが。

鈴木了二[SUZUKI Ryoji] 184▶197

杉並の住宅 物質試行13[Experiences in Material 13]

1. 東京都杉並区
2. 高増寛治
3. 夫婦２人
4. a. 鈴木了二建築計画事務所 鈴木了二、岡北良三、桜田滋 b. 山辺建築構造設計事務所 山辺豊彦、澤里敏明 c. 鈴木了二建築計画事務所
5. 建築：葛工務店、 設備：松本設備 電気：坪井電気
6. a. 180.47㎡ b. 89.64㎡ c. 154.07㎡
7. RC造
8. RC
9. 壁：RC、PB厚12EP、榀ベニヤ厚6-OP、京ジュラク塗、ドロマイトプラスター塗 床：玄晶石貼り、ブナ縁甲板貼り、モルタル金コテ押エ、Pタイル貼り、タタミ敷込み 天井：PB厚9-EP、榀ベニヤ厚6目透し貼り、そり板厚25打込み-EP

湯沸器：ノーリツGT-16 冷暖房機器：日立 厨房機器：サンウェーブ 洗面器、バスタブ、便器：TOTO 照明器具：ヤマギワ・ナショナル 建具：不二サッシュ

A商店ビル 物質試行21[Experiences in Material 21]

1. 東京都港区
2. 青島賢治
3. 夫婦２人＋子供２人＋老夫婦２人
4. a. 鈴木了二建築計画事務所 鈴木了二、桜田滋、箕原真、上村均 b. 山辺建築構造設計事務所 山辺豊彦、澤里敏明 c. 鈴木了二建築計画事務所
5. 建築：バウ建設 設備：原田設備 電気：フジ電気
6. a. 87.07㎡ b. 72.959㎡ c. 294.394㎡
7. RC造
8. RC、アルミパネル アルマイト処理900×900、
9. 壁：RC、PB厚12-EP、榀ベニヤ厚6-OP、アルミパネル厚5-アルマイト処理 床：黒御影石貼り、フローリング貼り、コルクタイル貼り、モルタル金コテ押エ、チェッカーP厚8、タタミ敷込み 天井：コンクリート打放し仕上げ、そり板打込み厚30-EP吹付け、PB厚12-EP
10. 洗面器、バスタブ：TOTO 冷暖房機器：ダイキン、ナショナル 照明器具：ヤマギワ、マックス 建具：昭和鋼機
11. 仮に、建築を「異質なもののせめぎ合う場」と定義する。そこでは「ディテール」とは、異質な諸要素が相互になんらかの作用をもつことによって出現する特異点をさす。その前提に立って物質試行No.13とNo.21の「ディテール」を比較するならば、それらがきわめて対照的位置にあったことがわかる。その決定的相違は、異質性の枠組の相違である。No.13では、それは線、面、マッス、グリッド等の全体に発散しうるシステムのレベルにあり、No.21にあっては、素材、色彩、幾何学的要素等の発散しえない断片的個別性のレベルにある。No.13では、それぞれかなり強力な運動量をほぼ等しく保有するシステム(たとえばコンクリートのR壁、木軸のグリッド、斜めの鋼材)が互いに抑制し合う傾向をもち、No.21では、運動量をほとんどもたず、いわば基底状態にある単純で静的なコンクリートの軀体に対して、さほど強力ではないが多種の動的要因(たとえば、４つの素材、２色の原色、10通りのR)が作用し励起し合う傾向をもつ。No.13では、「ディテール」はシステムの通過点に発生し、その複合性は不可視の領域(たとえば型枠あるいは軸線、通り芯)にあり、視覚的には単一であり、異質なるものの残した痕跡であり、したがってそれぞれのシステムの内部に属している。一方、No.21においては、複数の異質要素が出会いぶつかり合う境界に「ディテール」は生起しており、複合性は可視的というよりも振動的(色力と物性と形質との相互作用)であって、おのおのの接線上、あるいは表面上にあり、いずれにせよ諸要素の外部に「ディテール」は常に属することになろう。

高松 伸[TAKAMATSU Shin] 198▶213

修学院の家II[House in Shugakuin PartII]

1. 京都市左京区
2. 木下茂森
4. a. 高松伸建築設計事務所 b. 山本・橘建築事務所 c. 山京エンジニアリング
5. 山田工務店
6. a. 266.67m² b. 131.94m² c. 207.11m²
7. 鉄筋コンクリート造
8. 屋根：一般屋根 シート防水、ヴォールト屋根スチールパネル加工レタン塗装 外壁：コンクリート打放し、スチールパネル加工レタン塗装
9. 玄関：床 テラゾン張り、壁 コンクリート打放し、天井 プラスターボード厚12ジョイント工法VP ホール：床 テラゾン張り、壁 モルタル金鏝押え、プラスターボード厚12ジョイント工法VP
11. 素材には形や大きさがあるとおもう。というより形や大きさには固有の素材を選び寄せる力が備わっていると逆説めいていった方が良いのかもしれない。ともあれ私には、形が生み出す力を過度に信じ込んでいる節があり、その一種の力学に最も重要な因子は素材であると思っている。この際少々乱暴に言い切ってしまうとしたら、「素材は形に従う」などというきどった言い方になる。いつも様々なわけの分からぬ形のスケッチから設計を始める。紙の上を埋めてゆくちりぢりの形は、しかし私の場合常に重さや手触りや硬さなどをもった確たる素材でもある。私の設計にいつもつきまとう内向けの窮屈さは、そんな避け難い私自身の目と手の習性であるということを、すっかりと解ってしまっているくらいそれは私にはどうしようもないことなのだ。優しくはあるがどこかキリリと冷ややかな形の秀作であるこの住宅では、始めから終わりまでコンクリートの温度のようなものが気になっていた。いま、音も言葉も一瞬にして冷えてしまう空間に興味がある。

宮田邸[Miyata House]

1. 京都市北区
2. 宮田正道
3. 3人
4. a. 高松伸建築設計事務所 b. 山本・橘建築事務所 c. 建築環境研究所
5. 山鉄ハウス工業
6. a. 198.36m² b. 113.63m² c. 279.23m²
7. 鉄筋コンクリート造
8. 屋根：陸屋根部分：均しモルタル、樹脂系シート 防水、防水モルタル、ヴォールト屋根部分：鉄骨下地、スチール厚2.3曲げ加工防錆処理レタン塗装
外壁：RC、一部御影石張り磨上げ
外部建具：スチールサッシ、防錆処理、レタン塗装
9. 1階ガレージ：天井・壁 RC、床 モルタル金鏝押え
1階トレーニングルーム：天井・壁 RC、床 コルクタイル張り
2階居間・書斎：天井・壁 プラスターボードVP仕上げ、床 カーペット敷き
3階寝室・子供室：天井・壁 ラワンベニヤ二重張りVP 床 カーペット敷
3階ホール：壁 モルタル金鏝押えVP 床 ナラフローリング
11. 素材そのものを営々と鍛えあげることによって完遂される精度の美学や、こまやかで辛抱強い情愛によってのみ紡ぎ出される素材のエッセンスの芳香への興味はつきないのだけれど、どちらかというといつでも、素材と素材の甘美な関係の方にこだわってしまう。したがってたとえばコンクリートのような、その都度気儘に風味を異にする移り気な素材に対しても、その素材性を強引に定義してしまうような力業よりも、それ以外の素材との触れ合いや掛け引きによって生まれる妖しい風情を楽しむところについつい身を置きながら仕事をしてしまうことになる。この住宅においても、いわゆるコンクリート打ち上がりの謹厳さや、パネルの残す数学的な規律へのこだわりよりも、むしろ石や金属の多分に装飾的な華やぎをゆったりと抱き取ってしまうような大きな壁を打つことの方がより重要であった。そういう意味ではこの全長20mの壁は少々荒くはあるがなかなかに腰が座っている。

富永 譲[TOMINAGA Yuzuru] 214▶221

魚志ん[Uoshin]

1. 東京都文京区
2. 小川隆男
3. 夫婦＋母親＋子供1人＋(関西割烹)
4. a. 富永譲＋フォルム・システム研究所 b. 松井源吾＋O.R.S事務所 c. 茜設備事務所
5. 鈴木組
6. a. 64.02m² b. 55.74m² c. 266.86m²
7. RC壁構造
8. RC、一部大理石貼り
9. RC、一部椙合板貼り、青ひば練付合板貼り、土壁、しっくい塗り等
11. 灰色で平滑で冷たいコンクリートの表面と異なった質感の材料を同一面に並置することにより単一な厳しい感じを柔げようと試みた。
住居では椙合板。
店では青ひば板(黄金色)、土壁(黄金色)、
みがきしっくい壁(白色)、大理石(白色)、ステンレス板(灰色)、うるし(灰色)のカウンター、アルポリック板(灰色)、

渡辺豊和[WATANABE Toyokazu] 222▶241

オッパイハウス(戸口邸)[The Breast House]

1. 大阪府枚方市
2. 戸口勉
3. 夫婦＋子供2人
4. a. 渡辺豊和　b. 川崎建築構造研究所 川崎福則　c. 渡部豊和建築工房
5. 大藤建設
6. a. 199.24㎡　b. 80.93㎡　c. 170.57㎡
7. 鉄筋コンクリート壁式構造、ドーム部構造鉄骨造
8. 屋根：ヴォールト屋根：アイゾール防水(A工法)　外壁：RCドーム：アスファルトシングル葺き
9. 地階ホール・床：ノサヒフローリング　壁：RC、一面のみ陶壁画タイル貼り　天井：RC上VP吹付け
10. 湯沸器：LY-507(ターダ)　厨房機器：タカラ　便器、洗面器、バスタブ：TOTO

神殿住居 地球庵[Divine House "Chikyuan"]

1. 兵庫県西宮市
2. 藤田邦彦
3. 夫婦＋子供3人
4. a. 渡辺豊和　b. 川崎建築構造研究所 川崎福則　c. 渡辺豊和建築工房
5. 豊建設株式会社
6. a. 500㎡　b. 90.20㎡　c. 173.24㎡
7. 鉄筋コンクリート壁式構造
8. 屋根：ウオタイト防水の上モルタル鏝押工　外壁：RC
9. 床：ナラフローリング厚15OSW、壁：RC　天井：RC
10. 厨房機器：サンウェーブ　便器、洗面器　浴槽：TOTO
11. なぜ打放しか。まずは私の育った山口文象先生のＲＩＡの教育にある。近代主義の驍将であった先生は当然のごとく虚飾と排すという信念であられたから私もそのことを知らずのうちに身につけていった。しかし近代主義にまっ先に反旗をひるがえしたのも私であってみれば、なぜ打放しかは深刻な自問とならなければならないかもしれない。しかしそのような自問をしてみたことは実はないのであって、私にとって打放しはコンクリートの素材が何よりも好きなことにつきる。それも石に近いからというのではなく塑性が強いからであり、言ってみれば彫刻家の粘土のようなものである。したがってコンクリートに素肌がいいということになる。このコンクリートの塑性を巧みに生かしたのが晩年のコルビュジエであるが、そのころの彼はもはや「近代主義者」ではなくなっていた。私自身の建築もこの晩年のコルビュジエから始まったし、したがってそれは素面仕上の打放しである。

山本理顕[YAMAMOTO Riken] 242▶261

GAZEBO

1. 横浜市神奈川区
2. 山本とし子
3. 祖母＋伯母＋夫婦(計4人)
4. a. 山本理顕設計工場　b. 小島建築構造事務所
5. 青木建設
6. a. 229.03㎡　b. 210.58㎡　c. 664.60㎡
7. 鉄筋コンクリート＋鉄骨造
8. 屋根：コンクリート直押工(タケイ式進化コンクリート防水)、デッキプレート＋シンダーコクリート＋塗布防水　外壁：RC(ウレタン樹脂仮枠)
9. 床：ナラ縁甲板OP拭取り、クリアウレタン　壁：プラスターボート寒冷紗　VP天井：プラスターボード　寒冷紗VP
10. 空調：ヒートポンプ　給湯：ガス湯沸器

ROTUNDA

1. 横浜市神奈川区
2. 茂本みさを
3. 祖母＋夫婦＋子供3人(計6人)
4. a. 山本理顕設計工場　b. コジマ設計　c. 国設備設計事務所
5. 松村組
6. a. 265.95㎡　b. 230.36㎡　c. 904.42㎡
7. 鉄筋コンクリート造＋鉄骨造
8. 屋根：コンクリート直押工(タケイ式進化コンタリート防水)・デッキプレート＋シンダーコンクリート＋塗布防水　外壁：RC＋アクリル・カラー・クリア
9. 床：ナラ縁甲板、VP拭取り、クリアウレタン　壁：プラスターボード　寒冷紗、VP天井：プラスターボード　寒冷紗、VP
10. 空調：ヒートポンプ　給油：ガス湯沸器
11. 〈GAZEBO〉も〈ROTUNDA〉も構法として見るなら、全く同じ類の建築である。ともに、RCラーメン構造の軀体の上に鉄骨造の軽い上屋が乗り、さらにその上に天蓋のようなふわりとした屋根がかぶさっている。
地形のようなRCの塊に鉄骨の柱や梁がピンでとめられている。そんなイメージである。RCと鉄骨との絡み合い方だけを注意しながらつくっていった。だからもし、2つの建築に多少の違いがあるとすれば、この絡み合い方の違いである。そこの違いが、この2つの建築を少しは違うものにしているのではないかと思う。
2つの建築は、ちょうど1年ほどの時間差で完成している。〈ROTUNDA〉のほうが後からできた。そのぶん余裕がある。〈GAZEBO〉をおそるおそるつくった後、それとほとんど同じ手法ですぐまたつくる、というのは、1回目に比べれば多少は楽ができる。楽になった分だけ、見栄を切る余裕があるというわけである。
〈GAZEBO〉の力骨はメッシュの内側にひっそりと隠されているのに対して〈ROTUNDA〉のそれは逆に外側にでてきて、ファサードの印象を決定的なものにしている。天蓋の形にしてもRC部分と鉄骨部分の絡み合い方にしても少しずつ、〈GAZEBO〉より〈ROTUNDA〉のほうが、見せることを意識して作っているようにもみえると思う。どちらが完成度が高いと言ったらいいのかよくしらないけれども、苦労したのは〈GAZEBO〉のほうで、その苦労はほとんど表面にでてこない。視覚化されていない。

吉田保夫[YOSHIDA Yasuo] 262▶270

田中邸[Tanaka House]
1. 大阪市福島区
2. 田中庸吉
3. 夫婦2人
4. a. 吉田保夫建築研究所　末松利紀　b. 的場構造設計事務所　的場正通
5. 吉田保夫、末松利紀、藤原伸治、
6. a. 125.00㎡　b. 72.00㎡　c. 148.00㎡
7. 鉄筋コンクリート造2階建
8. RC
9. RC
11. 戦後に建てられたとおもわれる老朽化した建物が密集し、ひしめきあう家の音が聞こえてきそうである。擦れ違う人達の表情に覇気はなく、老人が1人で歩いているのをよく見かける。乾いた町である……。
ここに私達に課せられた敷地はある。道路らしきものはなく、ただ820mmほどの隙間が街とのコンタクトを可能にしていた。敷地が挑発しているように思えた。何十年もの間ここを守り続けた老人の意志もあった。いろんな問題が頭をよぎった。しかし、計画は進むこととなった。
2階を住居として使用、あたりに景色といえるものはなく、窓は最上部と最下部に設けた、ただ1箇所、1,200角程の開口がこの住宅と現実とを結んでいる。かろうじてタラップでつながっている1階には光を望むことはできず、床も壁も梁もコンクリートのまま、地階の様相を漂わしており、コンクリートの建具によって外界と遮断されている。
ざらざらとしたコンクリートの壁が家並みを横切り、空から見るとこの建築は都市を挑発する刃物のように見える。
(末松利紀)

吉田邸[Yoshida House]
1. 大阪府四條畷市
2. 吉田晶二
3. 夫婦＋子供2人
4. a. 吉田保夫建築研究所　b. 的場正通構造研究所
5. 吉田保夫、川原治、末松利紀、藤原伸治、阪口雅嗣、TORBEN BERNS、平川雅治、清水由紀、河北みちる、西口浩美、林彰伸、柳尾浩、岩本次幸、
6. a. 113.30㎡　b. 67.37㎡　c. 138.28㎡
7. 鉄筋コンクリート造3階建
8. RC
9. RC
11. 3度目の冬である。随分長い間この現場には引っ張られていたように思う、この間私を襲い続けた、不安・焦り・怒り・孤独・愛着などの表情が凝縮しているためか、客観的に見ることができない。
建築がもちうる意味やその存在が安易にそてし希薄になり、「型」にすらインパクトを求めることすらできなくなってしまった現在、すべてのことを受け止めるときなのである。
建築が成立しうるまでのプロセスの一部を取り除き自らの手によって破綻をあらわにし、ぎりぎりのところで「物」を「型」に置き換え、「物」そのものの意味を浮上させることばかり考えていた。
この現場に関わった誰もが輝いていた、そして同じ夢を見続けていてほしいと思う。　(末松利紀)

[文責：編集部]

作家紹介[Biographies]

阿部 勤[ABE Tsutomu] ► 16…25
■born in 1936, Tokyo
■address : 1-12-17 Gohongi, Meguro-ku, Tokyo
■1936年東京生まれ／1960年早稲田大学建築科卒業
■1960～1971年坂倉準三建築研究所／1971年アーキヴィジョン設立／1975年アルテック建築研究所設立／東海大学・早稲田大学講師／現・日本大学芸術学部講師
■スタンレー電気技術研究所／賀川豊彦記念松沢資料館／蓼科レーネサイドスタンレー／五本木ハウス
■1985年日本建築家協会新人賞
■連絡先：東京都目黒区五本木1-12-17

安藤忠雄[ANDO Tadao] ► 26…47
■born in 1941, Osaka
■address : 2-5-23 Toyosaki, Oyodo-ku, Osaka city, Osaka
■1941年大阪生まれ
■1969年安藤忠雄建築研究所設立
■六甲の教会／住吉の長屋／六甲の集合住宅／中山邸／下町唐座／水の教会
■1979年日本建築学会賞／1983年日本文化デザイン賞／1985年アルヴァ・アアルト賞(フィンランド建築家協会)／1985年毎日デザイン賞／1986年芸術選奨文部大臣賞新人賞／1986年毎日芸術賞
■「現代の建築家」(鹿島出版会)／「日本の建築家」(丸善)
■連絡先：大阪府大阪市大淀区豊崎2-5-23

東 孝光[AZUMA Takamitsu] ► 48…59
■born in 1933, Osaka
■address : 3-6-1 Minamiaoyoma, Minato-ku, Tokyo
■1933年大阪生まれ／1957年大阪大学工学部構築工学科卒業
■1957年郵政省建築部／1960年坂倉準三建築研究所／1968年東孝光建築研究所設立／東京造形大学・東京電機大学・東京大学・大阪大学講師／1985年大阪大学工学部環境工学科教授
■塔の家／さつき保育園／ワットハウス／日本バプテスト教会連合センター／聖書キリスト教会会堂
■「日本人の建築空間」「続・現代建築の再構築」(彰国社)／「都市住居の空間構成」(鹿島出版会)
■連絡先：東京都港区南青山3-6-1

原 広司[HARA Hiroshi] ► 60…77
■born in 1936, Kanagawa
■address : 701 Sakuragaoka Highhome, 13-10 Sakuragaoka, Shibuya-ku, Tokyo
■1936年神奈川県生まれ／1959年東京大学工学部建築学科卒業／1964年東京大学数物系大学院建築学専攻博士修了
■1962年設計同人RAS設立／1964年東洋大学工学部建築学科助教授／1969年東京大学生産技術研究所助教授
■旭硝子渋谷社宅／国民宿舎丹沢ホーム／鳥羽海の博物館／末田美術館／田崎美術館／ヤマト・インターナショナル
■1987年日本建築学会作品賞／1988年村野藤吾賞
■「建築に何が可能か」(学芸書林)／「空間　機能から様相へ」「集落の旅」(岩波書店)
■連絡先：東京都渋谷区桜丘13-10桜丘ハイホーム701

長谷川逸子[HASEGAWA Itsuko] ► 78…87
■born in Shizuoka
■address : 401, 1-9-7 Yushima, Bunkyo-ku, Tokyo
■静岡県生まれ／関東学院大学建築学科卒業
■1964～1969年菊竹清訓建築設計事務所勤務／1969～1971年東京工業大学建築学科研究生／1972年女子美術大学講師／1979年長谷川逸子・建築計画工房設立
■眉山ホール／練馬の家／東玉川の家／藤沢市湘南台文化センター
■1986年日本建築学会作品賞／1986年日本文化デザイン賞
■「住民時代」(新建築社)
■連絡先：東京都文京区湯島1-9-7-401

早川邦彦[HAYAKAWA Kunihiko] ► 88…101
■born in 1941,Tokyo
■address : 5F NW House, 1-3-7 Nishiwaseda, Shinjuku-ku, Tokyo
■1941年東京生まれ／1966年早稲田大学理工学部建築学科卒業／1971年イェール大学建築芸術学部大学院卒業
■1966年竹中工務店東京支店設計部／1978年早川邦彦建築研究室設立／その後、関東学院大学・早稲田大学・東京工業大学・工学院大学・武蔵野美術大学・東京大学講師
■成城の家三部作／中澤ホール／アトリウム／ANGLE
■1985年日本建築家協会新人賞
■「現代建築／空間と方法　早川邦彦」(同朋舎出版)
■連絡先：東京都新宿区西早稲田1-3-7NWハウス5F

伊東豊雄[ITO Toyo] ► 102…109
■born in 1941, Seoul (Republic of korea)
■address : Fujiya bldg. 1-19-4 Shibuya, Shibuya-ku, Tokyo
■1941年京城生まれ／1965年東京大学工学部建築学科卒業
■1965～1969年菊竹清訓建築設計事務所勤務／1971年URBOT(アーバンロボット)設立／1979年伊東豊雄建築設計事務所に改称／現・早稲田大学理工学部建築学科非常勤講師
■笠間の家／シルバーハット／ノマド
■1984年日本建築家協会新人賞／1986年日本建築学会賞作品賞
■「現代建築／空間と方法ー伊東豊雄」(同朋舎出版)
■連絡先：東京都渋谷区渋谷1-19-4不二屋ビル

石井和紘[ISHII Kazuhiro] ► 110…117
■born in 1944, Tokyo
■address : Studio Akasakalo, 4-14-27 Akasaka, Minato-ku, Tokyo
■1944年東京生まれ／1967年東京大学工学部建築学科卒業／1969東京大学工学系大学院修士修了／1975年イェール大学建築学部修士修了／1975年東京大学工学系大学院博士修了
■1976年石井和紘設計研究所設立／早稲田大学国際部・UCLA大学・日本大学理工学部・イェール大学建築学部講師／現・大阪大学工学部建築学科講師
■54の窓／バイ・コースタルハウス／田辺エージェンシー本社ビル／54の柱／ジャイロ・ルーフ
■連絡先：東京都港区赤坂4-14-27スタジオ赤坂楼

石井 修[ISHII Osamu] ► 118…129
■born in 1922, Nara
■address : Biken-Architecture Designing Office, 3-1-39 Fukushima, Fukushima-ku Osaka city, Osaka
■1922奈良県生まれ／1940年吉野工業学校建築科卒業
■大林組勤務／1956年美建・設計事務所設立
■豊国ガラスステーション／ドムス淀川／シャルル本社ビル／カトリック司祭の家
■1976年商業空間デザイン賞／1983年岡山市優秀建築物賞／1985年奈良都市景観調和デザイン賞、神戸市建築文化賞、兵庫県みどりの建築賞
■「家家」(学芸出版)
■連絡先：大阪府大阪市福島区福島3-1-39美建・設計事務所

槇 文彦[MAKI Fumihiko] ► 130…137
■born in 1928, Tokyo
■address : Green bldg., 3-6-2 Nihonbashi, Chuo-ku, Tokyo
■1928年東京生まれ／1952年東京大学工学部建築学科卒業／1954年ハーバード大学修士修了
■ワシントン大学・ハーバード大学準教授／1965年槇総合計画事務所設立／ハーバード大学・カリフォルニア大学・ウィーン工科大学客員教授／1979年東京大学工学部建築学科教授
■大阪府臨海スポーツセンター／沖縄国際海洋博政府出展水族館／岩崎美術館／電通大阪ビル／スパイラル／代官山ヒルサイドテラス
■連絡先：東京都中央区日本橋3-6-2グリーンビル

村上 徹[MURAKAMI Toru] ► 138…143
■born in 1949, Ehime
■address : 405, 2-1-4 Kogokita, Nishi-ku, Hiroshima city, Hiroshima
■1949年愛媛県生まれ/1972年広島工業大学卒業
■鈴ヶ台の家/己斐の家/黒瀬の家/大野の家/本郷町の家
■連絡先:広島県広島市西区庚午北2-1-4-405

室伏次郎[MUROFUSHI Jiro] ► 144…157
■born in 1940, Tokyo
■address : 1-12-17 Gohongi, Meguro-ku, Tokyo
■1940年東京生まれ/早稲田大学理工学部建築学科卒業
■1963年坂倉準三建築研究所/1971年建築研究所アーキヴィジョン設立/1975年アルテック建設研究所設立/1985年神奈川大学非常勤講師
■大和町の家/国立ハウス/岸病院
■連絡先:東京都目黒区五本木1-12-17

内藤 廣[NAITO Hiroshi] ► 158…165
■born in 1950, Kanagawa
■address : 301 Matsuoka Kudan bldg., 2-2-8 Kudanminami, Chiyoda-ku, Tokyo
■1950年神奈川県生まれ/1976年早稲田大学大学院修了
■1976年フェルナンド・イゲーラス事務所勤務/1980年内藤廣建築設計事務所設立
■TOM/静棲住居
■連絡先:東京都千代田区九段南2-2-8松岡九段ビル301

六角鬼丈[ROKKAKU Kijo] ► 166…177
■born in 1941, Tokyo
■address : Daiichi Yazaki bldg., 3-62-5 Wada, Suginami-ku, Tokyo
■1941年東京生まれ/1965年東京芸術大学美術学部建築学科卒業
■磯崎新アトリエ/1969年六角鬼丈計画工房設立/東京芸術大学講師
■八卦ハウス/雑創の森学園/金光教福岡高宮教会/大雪展望塔/都立武道館
■1979年吉田五十八賞
■連絡先:東京都杉並区和田3-62-5第一矢崎ビル

鈴木 恂[SUZUKI Makoto] ► 178…183
■born in 1935, Hokkaido
■address : 2-31-1 Jingumae, Shibuya-ki, Tokyo
■1935年北海道生まれ/1959年早稲田大学理工学部建築学科卒業/1959~1962年早稲田大学吉阪隆正研究室所属/1962年早稲田大学大学院修士修了
■1964年鈴木恂建築研究所設立/1980年早稲田大学教授
■ATL/PEH/大阪EXPO大噴水/雲洞庵仏舎利堂/龍谷寺妙光堂
■1971年C.R.ヴィリャヌエバ国際賞最優秀賞
■「太陽がいっぱいの家」(実業之日本社)/「メキシコ・スケッチ」(丸善)/「住居の構想」(鹿島出版会)
■連絡先:東京都渋谷区神宮前2-31-1

鈴木了二[SUZUKI Ryoji] ► 184…197
■born in 1944, Tokyo
■address : 508, 1-1-20 Azabudai, Minato-ku, Tokyo
■1944年東京生まれ/1968年早稲田大学理工学部建築学科卒業/1974~1977年早稲田大学池原研究室
■1970年fromnow結成/1977年fromnow建築計画事務所設立/1982年鈴木了二建築計画事務所に改称/現・早稲田大学・東京芸術専門学校(TSA)講師
■物質試行/断層建築/ブレヒト「ガリレイの生涯」
■連絡先:東京都港区麻布台1-1-20-508

高松 伸[TAKAMATSU Shin] ► 198…213
■born in 1948, Shimane
■address : 36-4 Takeda Jobodaiin-cho, Fushimi-ku, Kyoto city, Kyoto
■1948年島根県生まれ/1979年京都大学工学部建築学科博士修了
■1980年高松伸建築設計事務所設立/精華大学美術学科・京都工芸繊維大学建築学科講師
■先斗町のお茶屋/西福寺/織陣/ARK/PHARAOH/KIRIN PLAZA
■1984年日本建築家協会新人賞/1985年ビエンナーレ入賞(ヴェネチア)/1987年商空間デザイン特別賞
■「WORKS 高松伸」(グラフィック社)/「現代建築/空間と方法　高松伸」(同朋舎出版)/「僕は時計職人のように」(住まいの図書館出版局)
■連絡先:京都府京都市伏見区竹田浄菩提院町36-4

富永 譲[TOMINAGA Yuzuru] ► 214…221
■born in 1943, Taipei (Taiwan)
■address : 611, 4-12-16 Hongo, Bunkyo-ku, Tokyo
■1943年台北生まれ/1967年東京大学建築学科卒業
■1967~1972年菊竹清訓建築設計事務所勤務/1972年富永譲+フォルムシステム研究所設立/1973~1979年東京大学建築学科助手/日本女子大学・東京大学建築学科講師
■新ゆりグリーンタウン高層集合住宅/特別養護老人ホーム「幸風苑」ディサービス併設
■「近代建築の空間再読」(彰国社)/「現代建築/空間と方法　富永譲」(同朋舎出版)
■連絡先:東京都文京区本郷4-12-16-611

渡辺豊和[WATANABE Toyokazu] ► 222…241
■born in 1938, Akita
■address : 4F Shibakawa bldg., 4-33 Fushimi-cho, Higashi-ku, Osaka city, Osaka
■1938年秋田県生まれ/1961年福井大学工学部建築学科卒業
■1970年渡辺豊和建築工房主宰/1986年京都芸術短期大学教授
■吉岡邸/西脇市立古窯館/龍神村民体育館
■1988年日本建築学会賞
■「縄文夢通信」(徳間書店)/「天の建築、地の住居」(人文書院)/「建築を侮蔑せよ　さらば滅びん」(彰国社)
■連絡先:大阪府大阪市東区伏見町4-33芝川ビル4F

山本理顕[YAMAMOTO Riken] ► 242…261
■born in 1945, Peking (People's Republic of China)
■address : A-403 Twin-Building Daikanyama, 30-3 Sarugaku-cho, Shibuya-ku, Tokyo
■1945年北京生まれ/1965年日本大学理工学部建築学科卒業/1971年東京芸術大学大学院美術研究科建築専攻修了
■1973年山本理顕設計工場設立
■山川山荘/石井邸/山本邸/藤井邸
■1975年鹿島賞/1985年日本建築学会賞作品賞
■連絡先:東京都渋谷区猿楽町30-3ツインビル代官山A棟403

吉田保夫[YOSHIDA Yasuo] ► 262…270
■born in 1947, Saga
■address : 15-11 Karahori-cho, Tennoji-ku, Osaka city, Osaka
■1947年佐賀県生まれ/1969年大阪工業大学第二工学部建築学科中退
■1971年安藤忠雄建築研究所/1974年吉田保夫建築研究所設立
■多木邸/サンジェルマン
■1969年万国博国際共同館コンペ最優秀賞/1980年久保田鉄工・新建築社主催建築写真コンペ二席/1985年SDレビュー入賞/1986年SDレビュー大阪賞
■連絡先:大阪府大阪市天王寺区空堀町15-11

あとがき

■「新建築」誌の1982年2月号から11月号にかけて、内田祥哉が4回にわたって書いた時評コラム「コンクリート建築は歴史的建造物となり得るか」で、コンクリート建築は果して100年の耐久力をもつか疑わしい、とくに最近のものほどむしろ脆くなっている傾向がある、等々が指摘された。ごく短い文章であるにかかわらず、新聞で大きくとりあげられたりして思いがけないほどの反響を呼び、それは現在にまで尾を引いている。例えば同じ欄で、その4年後に杉山英男は「ポスト・RC造」と題して、コンクリート建築の耐震性に根本的な疑問を投げかけ、今世紀初頭から現在まで、多くの研究者がこの問題に取り組んできたが、コンクリート造の『業』、『宿命』を克服できるのだろうか」、かつて日本が「木造を捨て、煉瓦造を捨ててきた大胆な決断が今また必要とされるのではないか」とまで論じている。また例えば、この提言をさらに補うかたちで、「住宅建築」の1988年8月号で杉山と高橋靗一が「問題としての〈RC〉をめぐって」について対談し、とくに高橋は、現代の都市においてはコンクリートは耐火のための被覆材でなく、むしろ鉄の内側でこそ強い充填材として使うべきだと話している。

■これはジャーナリズムの表面にあらわれたいくつかの現象を拾ったにすぎない。だれにとっても100パーセント完全ではない素材に対する、同様な気持はあったにちがいない。コンクリート造のなかでも打放しコンクリートはとくに問題が多いのである。それは耐火耐震といった性能面からだけではなく、とくに一般の人々にとっては、日本の風土や町にはどうしてもそぐわない反建築素材として、長年親しまれて(！)きたのだ。

■この本のあとがきとしては、あまり穏やかではない話になってしまったが、そうした問題意識を踏まえたところで、にもかかわらず、日本の建築家たちが打放しコンクリートに託してきた、幅広く奥行の深い建築表現をできる限り克明に記述しておきたい。万が一、コンクリートの性能が全面否定されるにしても、現代建築が経験してきた、文化的記憶にさえなっているこの表現の圧倒的なキャパシティは、どこかにシフトされなければ、打放しコンクリートはいつまでも新たな、より深層的な様相の発生装置として生き残らざるを得ないのである。その生命力は、ある意味では木造や煉瓦造や鉄骨造より、まだはるかに強いのだ。

■ここに紹介する21人の建築家は、こうした多層的な表現の現在を示している代表として選んだ。同時に、筆者がその建築活動のほとんどを実際に見てきた建築家に限っているともいえる。ひとりひとりの紹介に際しては、この本に収録されていない過去の仕事に触れている場合もあり、読者の皆さんに対しての不親切を承知で、全体的な脈絡をとらえたかった。

■収録作品は、すべて新たに撮影された。住む人や日常の品々が画面に見えなくても、竣工写真とは違う、住宅の現在の姿が感じられれば幸いである。そのために大変な無理をお願いした住人の皆さん、建築家諸氏に厚くお礼を申し上げたい。

■本の制作に直接関わったのは、写真家の下村純一さん、デザイナーの田村祐介さん、企画・編集のグラフィック社・赤平覚三さんと私である。出来上がったものはさっぱりしているが、面倒な作業が長く続いて、だれがいちばん苦労したか分からないほどだけれど、私でないことは間違いない。これまで私が手掛けてきた雑誌や本のほとんどは、自分のアート・ディレクションやレイアウトでまとめてきたし、写真のアングルを指定することさえあった。結局、自分ひとりの好みや考えのうちにおさめてしまう傾向が強かったわけだが、ここでは新しい刺激を体験した。とりわけ、赤平さんの提案された《ジャパン・ハウス》というタイトルは、今もある異和感がつきまとっているが、同時に自分では得られない客観性を目の前につき出された気がしている。

■打放しコンクリートは、日本の現代住宅史の展開にとって、迂回だったのか、それとも加速的な要因となったのか、その重要なポイントには、本書では言及していない。次にはこの問題を論ずるべきだろう。

1988年10月　植田 実

Afterword

■In the magazine "Shinkenchiku (New Architecture)", from the February issue to November issue of 1982, in a series of Mr. Yoshichika Uchida's four columns on current issues which was titled "Can concrete structures be historical constructions?" various problems were pointed out, such as the doubt that the durability of concrete structure may not be for 100 years, especially, recent constructions tending to become rather more fragile. Although those were only short comments, the matter was unexpectedly taken up in newspapers and much talked about, ever up till now. For instance, 4 years later in the same column, titled "Post RC Structure", Mr. Hideo Sugiyama voiced the basic doubt about the earthquake-resistance of concrete structures, saying that in spite of many researcher's efforts from the beginning of the century up to now, it isn't known whether "the Destiny of concrete structure can be solved or not". Also, he points out, in Japan, "we had to abandon wooden structures and brick buildings and this bold decision may neede to be charged now". Also to complement this suggestion in the August 1988 issue of "Residential Constructions", Mr. Sugiyama and Mr. Teiichi Takahashi talked about "RC as a problem", and Mr. Takahashi especially says that concrete is not the covering material for fireproofing in modern cities but the effective filling material to be used inside the steel.

■These are just a few examples taken superficially from journalism. There must have been similar impressions among the people about the material which can not be called 100% perfect. Among all types of concrete structures, bare ferroconcrete especially has many problems. Not only as regards their capacity of fire and earthquake proofing, but above all for people in general, it has been long recognized as the "Anti-construction material" which doesn't match the Japanese climate and town sceneries.

■As the postscript of this book, this has not turned out to be a very moderate talk. But since I am conscious of these problems, I would like to record, as faithfully as possible, the broad and profound architectural expressions in which Japanese architects have relied on ferroconcrete. Even it the special nature of concrete is denied absolutely, still, as the common experience shared by modern architectures, ferroconcrete will have to remain as the main medium for exdressing the underly its vitality of structures especially if the capacity of this expressions protruding from the others can not be shifted elsewhere. Its vitality is far stronger than those of wooden and steel structures.

■The 21 architects introduced here have been chosen as representing a wide range of visions. It can be said that these architects are limited to the extent that the author actually saw their architectural activities. In introducing the individuals, although there are other past works which are not included in this book, a logical construction as a whole was pursued.

■All the included works have been newly photographed. Even if the residents and the sense of life can not be seen in the picture, it is hoped that the real, actual form of house different from construction pictures may be felt. For that, I would like to thank the people who live in them and the architects who were kind enough to cooperate.

■Photographer Mr. Junichi Shimomura, Designer Mr. Yusuke Tamura, Planning and Editing Director Mr. Kakuzo Akahira of the Graphic-sha Publishing Co., Ltd. and myself. Much effort have collaborated myself. Much effort have collaborated in producing this book. I've used my own art directions and layout in compiling most of the magazines and books that I have been involved with, and in some occasions, I even designated photo angles. As the result, I was strongly inclined toward finishing the work according to my own taste and ideas. But here, I experienced an entirely new stimulus.

■The very important point, whether bare ferroconcrete has been a detour or an accelerating factor in the evolution of modern Japanese housing, is not taken of in this book. This matter will have to be discussed separately.

October 1988
UYEDA Makoto

［著者略歴］

植田実［UYEDA Makoto］
◇1935年東京に生まれる。
早稲田大学でフランス文学を専攻。
月刊《建築》編集部を経て、《都市住宅》編集長。同誌のアート・ディレクションによって、ADC賞受賞。
◇1976年よりADA EDITA, Tokyo 編集部にて多くの建築書を手がける。
1985年よりフリー。
◇現在、早稲田大学建築学科講師を経て多摩美術大学講師。住まいの図書館出版局編集長。建築評論家。

下村純一［SHIMOMURA Junichi］
◇1952年東京に生まれる。
早稲田大学で美術史学を専攻。
広告代理店を経てフリーの写真家に。
◇ヨーロッパを中心に建築写真を撮り続け、建築評論も各誌紙に発表。
◇現在、日本写真家協会会員。
◇写真集に《モダニズムの壁》(グラフィック社)、《近代の光》(求龍堂)。
◇著書に、《不思議な建築》(講談社)、他多数がある。

UYEDA Makoto
◇Born in Tokyo in 1935.
Studied French literature at Waseda University.
After working on the editorial staff of the monthly architecture magazine "Kentiku" Mr.Uyeda became chief editor of "Toshi Jutaku (urban housing)", where he received the ADC prize for his art direction.
◇From 1976, he had been part of the editorial department of A.D.A EDITA, Tokyo.
Since 1985, he has been doing related work on a freelance basis.
◇After lecturing at Waseda University's Department of Architecture, Mr. Uyeda presently works as a lecturer at the Tama Art University, chief editor of the SUMAI Library Publishing Company, and also as an architecture critic.

SHIMOMURA Junichi
◇Born in Tokyo in 1952.
Studied Art history at Waseda University.
After working for an advertising agency, Mr. Shimomura became a freelance photographer.
◇Following that, he has been photographing new and rare buildings mainly in Europe.
His ideas and comments on architecture have been presented in various magazines.
◇He is now a member of the Japan Professional Photographers Society.
◇His photo collection includes albums like "Meaning of Walls in Modern European Architecture" by Graphic-sha and "Kindai no Hikari (Light of Modern Times)" by the Kyuryudo publication company. Among his many books are "Fushigi na Kenchikn (Unusual Architectures)" by the Kodan-sha company.

ジャパン・ハウス：打放しコンクリート住宅の現在
Japan Houses in Ferroconcrete

1988年11月25日初版第1刷発行

定価	4,800円
編著者	植田実© 下村純一©
発行者	久世利郎
印刷所	錦明印刷株式会社
製本所	錦明印刷株式会社
写植	株式会社プロスタディオ 株式会社三山綜合システム
発行所	株式会社グラフィック社 〒102 東京都千代田区九段北1-9-12 電話03・263・4318 振替・東京3-114345 落丁・乱丁本はお取替え致します。 ISBN4-7661-0460-9 C3052 ¥4800E